ORGANISATION FOR ECONOMIC CO-OPERATION AND DEVELOPMENT

ORGANISATION FOR ECONOMIC CO-OPERATION AND DEVELOPMENT

Pursuant to Article 1 of the Convention signed in Paris on 14th December 1960, and which came into force on 30th September 1961, the Organisation for Economic Co-operation and Development (OECD) shall promote policies designed:

— to achieve the highest sustainable economic growth and employment and a rising standard of living in Member countries, while maintaining financial stability, and thus to contribute to the development of the world economy;

— to contribute to sound economic expansion in Member as well as non-member countries in the process of economic development; and

— to contribute to the expansion of world trade on a multilateral, non-discriminatory basis in accordance with international obligations.

The original Member countries of the OECD are Austria, Belgium, Canada, Denmark, France, Germany, Greece, Iceland, Ireland, Italy, Luxembourg, the Netherlands, Norway, Portugal, Spain, Sweden, Switzerland, Turkey, the United Kingdom and the United States. The following countries became Members subsequently through accession at the dates indicated hereafter: Japan (28th April 1964), Finland (28th January 1969), Australia (7th June 1971), New Zealand (29th May 1973) and Mexico (18th May 1994). The Commission of the European Communities takes part in the work of the OECD (Article 13 of the OECD Convention).

Publié en français sous le titre :

LA BIOTECHNOLOGIE POUR UN ENVIRONNEMENT PROPRE

PRÉVENTION, DÉTECTION, DÉPOLLUTION

Foreword

Biotechnology for a Clean Environment – Prevention, Detection, Remediation is the report of the Ad Hoc Group of Government Experts on Biotechnology for a Clean Environment (see list at the end of the report). This Group, set up by the OECD Committee for Scientific and Technological Policy (CSTP), met four times (11-12 June, 1991; 27-28 April 1992; 11-12 January 1993; 14-15 October 1993) under the Chairmanship of Dr. Mike Griffiths (United Kingdom).

Biotechnology for a Clean Environment follows a number of earlier OECD reports which, since 1982, have illuminated various policy aspects and sectors of application of biotechnology.

This report responds to a need, widely perceived in OECD Member countries, for a scientific review of biotechnology's potential contribution to the prevention, detection and remediation of environmental pollution, and thereby incidentally addresses the occasional misapprehension that the environmental implications of biotechnology are mainly a cause for concern.

The specialists involved in the preparation of this report were: Dr. Mike Griffiths who wrote Chapters 1-5 (State of the Art), with the advice and assistance of Dr. Ron Atlas (United States), and edited the rest of the report; Dr. Salomon Wald, of the OECD Secretariat, who wrote Chapter 6 (Economic Perspective) and the annex on government R&D initiatives, and generally co-ordinated the work, and Dr. Paul Hesselink (The Netherlands) who wrote Chapters 7-12 (Industrial Aspects) with the help of Mrs. Christien Enzing.

Thanks are also due to the many scientists who contributed to the report, particularly to the chapters on the State of the Art. They include the members of the Ad Hoc Group; participants in the Workshop on R&D Priorities and Policy Recommendations in Environmental Biotechnology, held in Weybridge on 3-4 June 1993 (see list attached to Annex of Part I), and other scientists, specifically Dr. Peter Barratt, Prof. Rita Colwell, Dr. Roger Dean, Prof. Geoffrey Hamer, Dr. Joseph Heijnen, Dr. Bobby Millar, Prof. Vivian Moses, Dr. Paul Norris, Prof. David Rawson, Dr. Tom Stephenson and Dr. Andrew Wheatley. The assistance of companies who participated in interviews or conducted searches for Chapters 7-12 is also gratefully acknowleged.

During the entire work for this report, close co-operation was maintained with the OECD Environment Directorate and with the CSTP's Group of National Experts on Safety in Biotechnology (GNE). It was agreed that this report should not attempt to cover safety aspects of bioremediation, so as to avoid possible duplication of the activities of the GNE. Scientific safety issues will be taken up, together with the more general scientific follow-up to the present report, by an OECD Bioremediation Workshop to be hosted by Japan on 27-30 November 1994 in Tokyo.

Particular thanks are due to the European Commission (DG XII), the Government of The Netherlands (Ministry of Economic Affairs), the Government of the United Kingdom (Department of Trade and Industry) and the Government of Japan (MITI) for their generous contributions to the financing of this activity.

The Committee reviewed the report by written procedure in February 1994, and recommended its publication on the responsibility of the Secretary-General. The report does not necessarily reflect the views of the OECD or of its Member governments. In addition, it must be pointed out that mention of industrial companies, trade names or commercial products or processes in this report, be it in the "Case Studies" or elsewhere, does not constitute an endorsement or recommendation by the Ad Hoc Group or the OECD.

Table of Contents

Part I

THE STATE OF THE ART

Part II

ECONOMIC AND INDUSTRIAL ASPECTS

Executive Summary

Biotechnology is moving into its third, and possibly most important domain. After health care and pharmaceuticals – the first major sector of application – and the agricultural and food applications, which followed soon after, the protection and restoration of the environment could become a priority goal of the life sciences and technologies.

Environmental biotechnology must be seen in the context of a future in which industrial technologies will have to be increasingly in harmony with the global material cycles of the biosphere. Industrial technologies of the future must be friendly to humans and the environment, and thus have four key characteristics:

- they should be based on renewable resources;
- they should use "mild" production processes;
- the resulting products and services should be environmentally compatible;
- any waste they generate should be recyclable.

This OECD report shows that biotechnology can provide a large number of technical options which may help achieve these goals. Biotechnology is certainly not the only technology to make and keep the environment clean, but it is an essential one, and its importance, in synergy with other tools, is growing.

Biotechnology is much broader than gene technology. It is the application of biological organisms, systems and processes to the provision of goods and services. Thus, the report does not focus particularly on genetically modified organisms, but it does report two widely held scientific convictions: first, that there is still a large, unknown and little exploited potential of naturally occurring organisms which can contribute to the restoration and maintenance of the environment; second, and not contradictory with the first, there may also be a need to use genetically modified organisms *in situ* to degrade the more recalcitrant pollutants, since the natural evolutionary processes may be too slow. The importance of safety in these, as in any other applications of biotechnology, is recognised but not specifically investigated, in order to avoid duplication of OECD's important parallel activities on biotechnology safety.

Biotechnology, as noted above, can be applied to many economic sectors. The current opportunities for an expanded role in the prevention, detection and remediation of environmental damage are primarily due to the increasing pervasiveness of pollution across all sectors and countries, but also to the improving relative cost-efficiency of biological clean-up methods as compared to the more traditional physical and chemical ones. In fact, biotechnology also consumes less energy and may produce less or at least more manageable waste than many of the more traditional techniques.

At present, the main use of biotechnology is to clean up or "remedy" pollution. Waste water clean-up was one of the first applications of biotechnology, followed by air and off-gas cleaning (biofilters). The focus of bioremediation is now increasingly shifting towards soil and solid waste, thereby raising complex scientific and technical questions related to the little-understood interactions of organisms with each other and with soil.

However, it must be emphasised that biotechnology is a future-oriented technology, and the top priority for environmental biotechnology in the future will therefore be to reduce and prevent damage. The potential of bioremediation must be seen not as a licence to continue polluting, but as a solution which inevitably has to receive priority at the present time, when countless polluted sites are calling for urgent remedial action.

There are a number of ways in which biotechnology can prevent or reduce local environmental damage. These include:

- added value processes, which convert a waste stream into useful products and thus facilitate international commerce in these new value added products and help restore natural balances;
- "end-of-pipe" processes, by which the waste stream is purified to the point where the products can be released without harm into the environment;

- development of new biomaterials, leading to the manufacture of materials with reduced environmental impact;
- new biological production processes, which generate more manageable waste.

In addition to prevention, detection and monitoring by biological methods are likely to play a more and more significant role in the future. These methods will assay levels of pollution, give early warning of pollution incidents, and allow for measurement and control of processes in industry and for environmental clean-up.

The long-term market potential for environmental biotechnologies is vast (expected to grow from $40 billion in the early 1990s to perhaps $75 billion by the year 2000), and impacts on industrial growth and employment could be significant. The report shows that in all OECD areas, hundreds of suppliers and thousands of potential user companies, both big and small, already invest, or plan to invest, in environmental biotechnology.

The biotechnology industry already makes important contributions to the protection and clean-up of the environment, to varying degrees according to the sector and the state of advancement of the available technologies. Whereas classical bioprocesses have been applied for decades, often on a large scale, process-integrated industrial biotechnology ("clean technology") is still in its infancy. Industry views this as a very promising technology for sustainable development.

While this industrial optimism is an encouraging sign, it is clear that government policy could be one of the most important factors in the future development of biotechnology for a clean environment. Governments will have to play a key role in encouraging and funding research and innovation in environmental biotechnology – areas which have been somewhat neglected in the past so that in many countries, only from 2 to 10 per cent of total government expenditures for biotechnology R&D are channelled into environmental biotechnology.

Many environmental biotechnologies are still operated on "black box" principles, and performance often lacks reliability and predictability. Their future development will depend on continuing R&D effort, and particularly on progress in basic front-line research in microbial physiology, genetics, and ecology and its translation into novel or improved process engineering. The report makes specific recommendations for R&D for the treatment of air, soil and land, solid wastes, waste water, and industrial effluents, as well as for improved monitoring and environmental assessment, which cuts across all other sectors.

Environmental biotechnologies may also address global environmental challenges, such as global warming, desertification, water shortages and the need to produce clean and renewable fuels, by fixation of atmospheric CO_2 and the development of salt-tolerant plants, for example. International R&D co-operation under government sponsorship would be a very appropriate approach to these challenges.

Beyond R&D government policies can affect environmental biotechnologies in many other ways. For example, in various countries, particularly in Europe, the introduction and application of environmental protection legislation has created a market for, and encouraged the development of, a competitive environmental biotechnology industry.

Environmental biotechnology brings together many disciplines, interacts with many other branches of science and engineering, and may be seen as a sector in which public and private initiatives may successfully be linked.

Introduction

Biotechnology for a Clean Environment – Prevention, Detection and Remediation is divided into two parts. Part I, "The State of the Art" reviews the scientific and technological options which biotechnology offers for the prevention, detection and remediation of air, water and soil pollution.

Part II, "Industrial and Economic Aspects", describes industry strategies and perspectives and presents an economic perspective on environmental biotechnology. An annex gives information on government R&D initiatives.

The use of micro-organisms in environmental processes goes back well into the last century, although these applications might better be regarded more as art than science. By drawing on a line at the time, in the late 1950s and early 1960s, when the structure and function of nucleic acids were being discovered, then it is possible to distinguish between the earlier traditional biotechnology and second generation biotechnology, which in part makes use of recombinant DNA technology.

The definition of biotechnology adopted here is "the application of biological organisms, systems, and processes to the production of goods and services". It must be stressed that biotechnology is wider than genetic engineering and draws heavily upon process technology, chemistry, and classical engineering. Much in the way of bioreactor design and the use of immobilised cells should also be called second generation biotechnology. The definition of many terms associated with modern biotechnology, especially its environmental aspects, will be found in the "Glossary and Abbreviations" at the end of the report.

The environment, as Einstein rightly said, "... is everything that isn't me". The consequence of this is that *all* biotechnology has a relationship with the environment. It is both cause and effect, problem and solution. In order, therefore, to produce a well-defined and useful document, this report is limited to those applications of biotechnology which are appropriate to improving the quality of the environment, specifically in the compartments of air, soil and water.

The State of the Art describes present-day technologies, all of which have been directed at resolving localised environmental problems – an industrial effluent or an area of contaminated land, for example. In the future, more global issues may be addressed, and in Chapter 4 some possible biological solutions to these challenges are described. For example, biological mechanisms exist for removing of greenhouse and acid rain gases, for producing environmentally acceptable energy sources and materials, and for resolving problems of water shortage and desertification. Many of these are of laboratory interest only at present but have the capability of becoming major technologies in future.

Other areas of biotechnology have environmental implications and may contribute to the reduction of pollution whether that is their main aim or not. Two of these, agro-food biotechnology and biomining are described in the appendices, but are not discussed in the main body of the report.

The examples used in the report have been taken from OECD countries. However, all of the processes described – amelioration of polluted land or improvement of drinking water, for example – may be used in less developed countries and particularly in tropical areas where higher ambient temperatures result in more rapid biological reaction rates.

While the whole report, and particularly the State of the Art, attempts to present the leading edge of the technology, it is not written primarily for scientists and engineers but rather for the wider audience of policy makers, administrators and industrialists and, of course, for all lay persons who have an interest in the subject. There is general agreement that the public has insufficient knowledge both of biotechnology and of the science of environmental problems; it is hoped that this report will go some way towards meeting this need.

Little is said in the report about the many physico-chemical processes used for environmental clean-up. This is not intended to imply that biotechnological options are inevitably the best even though (see Part II) they may well be relatively cost-effective in many instances. Most pollution control technologies are not biological in nature and biotechnology is but one of a cluster of related technologies that are applied to environmental clean-up.

This report is not, however, a textbook on remediation technology, but is specifically intended to illustrate the increasing role biotechnology will play. All applications of these technologies are multi-disciplinary in nature and while emphasis is placed on the bio-aspect, the report refers to the wider context as appropriate. The key message is that the most efficient techniques for remediation may well include the synergistic use of physical, chemical **and** biological approaches.

Within the report as a whole, some important issues such as pollution prevention or emerging global problems may seem to be treated only briefly, while others, such as bioremediation of soil and water, are covered more extensively. This reflects the present scientific and technological situation, which is the outcome of past R&D history, and the current short-term policy priorities and pressures. The R&D proposals are meant to suggest some correctives for this situation.

Also, the rather brief nature of the review of current government initiatives in support of R&D in environmental biotechnology does not imply that in this sector government is of secondary importance. The contrary is true. However, the data and international statistics for a comprehensive chapter on this crucial issue simply do not yet exist. What the few available statistics reveal is that the proportion of total public biotechnology R&D spent on protecting the environment is very small (in most countries a few per cent), when compared to other applications.

Biotechnology, in conjunction with other technologies, can provide remedies, but underpinning science is needed to improve and optimise its use. Much of what has been done by way of clean-up and remediation has been pragmatic in nature, and most processes in this area may best be described as "black boxes". Future R&D in environmental biotechnology will be directed at elucidating the underlying mechanisms so as to develop better systems and controls and optimise performance.

For practical engineering applications on an industrial scale, processes must be both reliable and predictable, and biological waste treatment and remediation will only achieve its true potential with more fundamental knowledge.

Part I, "The State of the Art", endeavours to make the current state of the biotechnology of environmental clean-up more readily acceptable to the general reader. One of the main difficulties here is that the area is changing very rapidly. An effort was made to strike a balance between informing and overwhelming the reader with scientific information; where appropriate, the more technical details have been relegated to the appendices at the end of Part I. Cases studies have been liberally used throughout to illustrate specific applications of the technology.

Chapter 1 sets the scene by describing traditional methods of waste treatment, the nature of the problems, and some of the background science, particularly the nature of some of the underlying biological processes and the use of recombinant DNA (r-DNA) technologies.

It is accepted, by both industry and governments, that there is a considerable inheritance of land and water pollution which will require remediation and furthermore, that industrial and agricultural activities will continue to produce waste for the foreseeable future. Although some pollutants are already treated in a very sophisticated manner, others are still discharged into rivers or the sea, dumped on land, or vented into the atmosphere.

Three environmental compartments can be distinguished. The traditional treatment technology for gaseous pollution (*air*) includes scrubbing, and biotechnological processes such as the use of peat and heather filterbeds, which have been particularly successful in reducing malodorous compounds to very low concentrations.

Water is indispensable for all forms of life. In spite of this, fresh water resources, particularly, have been grossly abused by industrialisation and the resultant changes in population distribution. As the production of food and industrial crops has intensified, the use of artificial fertilisers and pesticides has grown, and with this growth has come the inevitable rise in the pollution of water courses. Most available and foreseeable technologies for waste water treatment plants rely on biological activities of microbial communities. Trickling filters and activated sludge plants have been in use since the early decades of this century or earlier.

Essentially, there are two major categories of polluted *land* environments. Those that have been polluted by either spillage or leakage during production, handling or use of industrial material, and those that have been used as disposal sites for diverse chemical wastes. Polluting substances can percolate through the soil and reach groundwater and hence aquifers. Both chronic and acute situations exist. Land that has been polluted as a result of long-established industrial activity requires rehabilitation, and there is continuous low-level water pollution by agriculture and industry alike. There are also the massive accidental releases of pollutants, such as oil spills. Traditional techniques for treatment of contaminated land include ploughing to aerate (and stimulate biological activity) and "pump and treat", where groundwater is extracted, treated by physical or biological means and re-infiltrated.

Soils and sediments are also at the receiving end of substantial amounts of solid wastes, such as sewage sludges and compost. The conversion of organic waste to organic fertiliser is intrinsically valid, but the quality of the material being recycled is of critical importance. Disposal of municipal solid waste by land-filling has up to now been rudimentary, with little or no control of the biological processes. In view of the heterogeneity of this material, little improvement is possible unless there is separation at source.

Aerobic composting is inherently slow, because of the slow rate of diffusion of oxygen into the particles. Anaerobic composting proceeds at high rates but the process is susceptible to "souring", when the microbial populations are no longer in balance. Under current agricultural practices in the developed countries of the temperate zones, the resulting humic product has little economic value, but it has great potential for enhancing soil productivity in developing countries.

Industrial pollution occurs in all environmental compartments. Production wastes are represented by acidic gases, toxic organic vapours, solvents, chlorinated organic molecules, and metal salts, all of which may be found in aqueous and gaseous effluents and contaminating industrial sites. A significant number of synthetic compounds or xenobiotics, unrelated to natural compounds, persist and accumulate in the environment. Major areas of concern are pesticides, organic solvents, and polyaromatic and polychlorinated compounds such as the dioxins. Many dyes and colours and some modern detergents are also only poorly degraded by conventional waste treatment.

In addition to the organic pollutants there are a number of important inorganic pollutants. Nitrate pollution is often a consequence of excess fertiliser application, and phosphate may be derived from domestic and other detergents. Between them, they support algal growth (eutrophication). Toxic metal ions represent a major contaminant of municipal sewage and industrial effluent. Zinc, copper and nickel are phytotoxic, while cadmium, lead, and chromium are more mobile and there is a direct risk from the consumption of contaminated food. Sludge from sewage plants concentrates the metals from industrial waste to the point where it may no longer be spread safely on land.

There is some controversy about the possible use of genetically modified organisms (GMOs) in the environment, and a continuing scientific debate, including that within the OECD's work on biosafety, is taking place. While many naturally occurring organisms can be applied for environmental clean-up, it may also be necessary to use genetically modified micro-organisms to degrade the more recalcitrant pollutants since the natural evolutionary processes are relatively slow.

Although the bulk of biological processes use micro-organisms, the role of higher plants and animals should not be disregarded, particularly in environmental clean-up, where they may play an important role.

Chapter 2 describes the most recent technological developments under the headings *prevention*, *detection* and *remediation*, both at source (often referred to as *biotreatment*) and following dispersion of pollutants (*in situ treatment* or *bioremediation*).

The new biotechnologies for a clean environment have evolved against a background of traditional methods for waste treatment and in response to increasing environmental problems. Industry and academe are presently giving considerable thought to the development of so-called "clean technologies", those that minimise or, ideally, give rise to no environmental burden. The long-term aim should be to prevent the creation of polluting materials in the first place, and, in the case of industry, cleaner technology is a priority goal.

There are a number of ways biotechnology can prevent or reduce environmental damage. These include:

a) added-value processes, which convert a waste stream into useful products;
b) end of pipe processes, in which the waste stream is purified to the point where the products can be released without harm into the environment;
c) development of new biomaterials, the manufacture of materials with reduced environmental impact;
d) new biological processes, which generate less waste.

Conventional chemical processes generate large amounts of waste and by-products through the use of high temperatures, extremes of pressure, and a wide variety of highly reactive chemicals. In contrast, biotechnological processes generally occur at moderate temperatures and pressures. Enzymes or micro-organisms are highly selective, and the use of additional chemicals is minimal.

Detection and monitoring can play a significant role in a number of contexts, first to assay levels of pollution and give early warning of acute pollution incidents, and second to allow the measurement and control of industrial biotreatment processes whether they are in reactors or *in situ*. Many of the systems of measurement are still rather rudimentary, but combinations of biological and physico-chemical approaches are yielding new techniques that are sensitive but robust.

Biotreatment and bioremediation technologies are best subdivided by the nature of the environmental "compartment" being treated. Thus, in this chapter, these activities are subdivided into: air and off-gases, soil and land, solid wastes, and waste water and industrial effluents, the last including drinking water.

For *air and off-gases*, peat and compost beds are effective for the breakdown of odours and simple volatile organic chemicals and in many cases offers a simpler and cheaper alternative to chemical treatment. However, improved technologies are required to deal with many industrial off-gases and to overcome some of the shortcomings of the existing biological systems, which may be too slow or have short lifetimes because of the accumulation of by-products. Recently, there has been much improvement in filterbeds and biofilters through the use of synthetic substrates and selected organisms.

Soil and land treatment may be either *in situ* or *ex situ*. The former involves a number of techniques, both biological and non-biological, in which the soil is not disturbed. This may be essential in a busy city location, for example. *Ex situ* treatment requires the soil to be excavated and treated above ground, either in piles or in specialised reactors. These procedures are easier to control than underground treatments. Bioremediation of land is often cheaper than physical methods and its products are largely harmless. It is, however, more time-consuming, and can therefore tie up both land and capital. Technologies for soil include *ex situ* processes such as composting, soil banking, slurry reactors and *in situ* processes such as nutrient solution injection and bioventing (where air is supplied both for micro-organisms and as a carrier for volatile materials).

The simplest and preferred *in situ* processes involve indigenous organisms which can be stimulated through the addition of nutrients to obtain a satisfactory level of remediatory activity. Alternatively, organisms possessing specific biological potential can be added to the site. Some well-documented demonstration sites are needed in order to provide a stronger base for the development of such technologies in this area.

Solid waste treatment aims at converting waste into a safer, less toxic, more stable material which can be used or disposed of. Techniques include deposition in landfills, composting (which is an aerobic process) in open piles or bioreactors, and anaerobic digestion of solids in order to convert the organic content into usable methane. With source-separated solid organics, a number of biotechnological processes are being introduced, which can address up to 30 per cent of the total.

While the traditional aim of biotreatment of *waste water and industrial effluents* was to reduce organic matter, the increasing importance of industrial pollution now a greater need for processes that remove specific pollutants. Among these, nitrogen and phosphorus are priority targets, as are heavy metals and chlorinated compounds.

Both aerobic and anaerobic processes and fixed bed and suspended reactor systems are used for water treatment. The appropriate choice depends on a number of factors, including the quantity, concentration and nature of the pollutants and the area available for the plant to occupy. As target levels for pollutant concentrations are reduced, combinations of physical processes, such as adsorption with biological degradation, will increasingly be used. Such combinations, together with the selection of organisms to degrade specific compounds, will be used to treat the more recalcitrant compounds.

Waste water treatment plants currently in use are essentially operated on a "black box" principle and lack reliability. Enhanced performances for aqueous effluent treatment have largely been due to improvements in reactor configuration (*e.g.* tower reactor, air lift, fluidised bed, etc.). Individual processes need to be optimised and controlled, for example by adjusting influent quantity and/or quality and modifying the microbial community.

Oils spills, groundwater and aquifer contamination, and the recycling of waste water into drinking water, together with the appropriate biological clean-up methods in each case, are treated in subsections.

Chapter 3 addresses the need for reasearch and development in environmental biotechnology while appreciating that the need will not be exclusively for biotechnology but will requre integration with other scientific disciplines. R&D divides into two primary areas, fundamental research and bioprocess engineering, which in turn are subdivided as follows:

Fundamental research:

- molecular biology;
- microbial physiology and metabolism;
- microbial ecology and genetics;

Bioprocess engineering:

- scale-up problems;
- modelling and control theory;
- analysis and measurement;

- reaction kinetics and growth-limiting factors;
- process design and optimisation;
- bioavailability.

Certain proposals for discrete R&D activities are the recommendations of the Ad Hoc Group, and take into consideration the advice of an international panel of scientists. These specific R&D recommendations are as follows:

Organisms:

- Organisms and mixed cultures should be screened and selected for specific functional capacities (*e.g.* in waste treatment systems specifically defined organisms could replace current sludges with undefined populations of organisms).
- Organisms and their interactions in biocommunities, as well as communities in the environment should be studied.
- Organisms with novel and/or improved functional phenotypes should be developed either by recombinant DNA methods or through the selection of evolutionary mutants.
- Inocula with known functions should be developed.

Bioprocesses:

- Processes based on the use of combined biological and physico-chemical technique, *e.g.* membranes, are required in order to develop new bioreactor designs.
- Extensive and intensive processes are needed for the treatment of diffuse trace levels of pollutants.
- A better understanding is required of the nature and functioning of "black box" processes, *inter alia* carbon dioxide fixation, methane and hydrogen production, nitrogen and sulphur metabolism, and production of biopolymers.
- Modification of environmental conditions, *e.g.* nutrients, acidity, etc., to enhance desired *in situ* biological conditions requires investigation.
- Processes occurring at surfaces, which underpin many biotechnological processes, need considerable research (*e.g.* kinetics, immobilisation, bioavailability, etc.).
- Model systems, *e.g.* microcosms, should be developed to predict operational performance and transfer techniques from the laboratory.
- Designs for technical processes are required for treating specific waste and extracting value added products. Specific pollutants should be pretreated to avoid release.
- Processes aimed at prevention or treatment of waste should be directed at recirculation and added value (clean water and air are value added products).
- Investigations should be performed to stimulate a wider use of recovered products locally and also internationally; indeed environmental biotechnology should contribute to proper mass fluxes and global balances of nutrients and other materials by facilitating international commerce of the latter.
- Development of biological materials such as biopolymers, fuels and base chemicals to replace those produced from fossil fuels, is needed.
- Metals recovery by biological means should be improved, together with their subsequent recycling.

Ecology and ecosystems:

- Better understanding is needed for the restoration and sustainment of impacted ecosystems (for example, forests, lake and paddy field ecosystems).
- Mechanisms both of plant survival in arid conditions, and of salt tolerance need further research.
- Mapping of "hidden" microbiological diversity and development of new culturing techniques are needed to exploit natural resources better.

Measurement and monitoring:

- Measurement tools, procedures, and protocols for site assessment should be improved.
- Sensitive but robust instrumentation and diagnostics need to be developed to identify both biological and chemical hazards, such as whole cell biosensors for water quality.
- Reliable test procedures must be developed to assess the fate of anthropogenic chemicals in full-scale treatment systems.
- New assays for specific pollutants are needed, including those based on immuno-technology.
- Better techniques need to be developed for monitoring mixed populations of micro-organisms and for tracking organisms within ecosystems.

- "On-line" toxicity tests should be developed to assess whether water streams or soils are contaminated or decontaminated.
- Operational clean-up criteria need to be formulated in order to determine how clean is clean: the production of intermediates of degradation, levels of metal ions, etc.

Chapter 4 examines the future of environmental biotechnology, not in terms of the priorities already described but rather through a crystal ball to see what might be the consequences if some of the R&D is successful. It does not encompass all R&D but rather gives some example of likely outcomes, some in the near future and others somewhat more distant. The examples are grouped under the headings:

- *in situ* bioremediation;
- real-time monitoring and delivery systems;
- bioprocessing;
- emerging global challenges;
- production of clean fuels.

Chapter 5 provides the reader with a brief guide to further sources of information. There is a wealth of scientific literature, in the form both of journals and electronic databases on environmental biotechnology. The brief bibliography and the database examples expound only briefly on the information given in this report. A more detailed bibliography and list of relevant databases is available.

Part II, "Industrial and Economic Aspects", begins (Chapter 6) by placing environmental biotechnology in a wider economic perspective, thus bringing the characteristic features of this technology into sharper focus.

In general, biotechnology started as a "science-push" or science-driven technology, with initially very little "market-" or "demand-pull" and virtually no direct government R&D financing or guaranteed public markets. This has put biotechnology at a distinct disadvantage in comparison with other major generic technologies of the twentieth century.

Due to recent environmental protection legislation in a number of countries, market conditions for biotechnology are now changing rapidly. In these countries, governments are creating a powerful market-pull which other biotechnology sectors (health, agro-food) did not know in the past.

Economic theory has shown that the successful diffusion of a new generic technology depends largely upon five criteria:

i) a new range of technically improved products and processes;
ii) cost reductions for many of these;
iii) social and political acceptability;
iv) environmental acceptability;
v) pervasive effects throughout the economic system.

The industry survey, which follows, shows that environmental biotechnology satisfies the first two criteria very well. Regarding the third criterion, opinion surveys indicate that biotechnology is socially and politically well accepted when used to protect the environment. Environmental acceptability (the fourth) is not at present controversial, as genetically modified organisms are not yet used in large-scale bioremediation. Their possible introduction in the future, however, will be evaluated in the light of appropriate safety analysis and safety management concepts, and work on this has already begun.

Finally, environmental biotechnology pervades all economic sectors, probably more so than any other application, due to the ubiquitous spread of environmental pollution. It therefore meets the fifth criterion.

The chapters on industry strategies and perspectives indicate which of the many scientific and technological options are already industrially available, and which are being developed, considering the costs and competitive advantages.

The main basis of this part of the report is a number of interviews with industrial companies, small, medium and large, including both users and producers of environmental biotechnology. In addition, some findings are also based on literature and data bank surveys.

The overall picture shows a rapidly growing field with many technologies that are competitive, have already become indispensable to environmental protection and clean-up, and are engaging tens of thousands of companies across the OECD area. Environmental biotechnology has begun to affect the industrial competitiveness of OECD countries, has enormous future potential, and is likely to become in the next century a sector of major economic importance for industrialised countries.

Chapter 7 describes the products, processes and services currently available. The role of industry in developing biotechnologies for *pollution detection and monitoring* is very small, as physical and chemical methods are most prevalent in this area, and few relevant data could be found. In contrast, *pollution prevention* through process-integrated biotechnology is an important and growing field. However, as the technology is integrated into process development and design, it is hardly "visible" in the form of concrete products and services. Hence, while it is easy to assess markets for specific products such as biofilters, it is impossible to estimate the overall market for "clean technology". In general, "clean technology" is still in an early stage of development, and its diffusion in some sectors, particularly chemical industries, is very slow.

Many more data can be presented for clean-up and treatment/abatement biotechnologies in each of the major sectors: air/off-gases, water, soil, and solid waste.

Biotechnology processes to remove contaminants from air and off-gases (biofiltration, bioscrubbing, bio-trickling filtration) are applied in industries that generate odorous gases. A second generation of such biotechnologies to treat toxic pollutants, and higher concentrations of pollutants, will soon reach the market.

The competitive advantages of biotechnology are, among others, a complete conversion of contaminants into harmless substances, cost reductions of 60 to 80 per cent compared to thermal and chemical processes, and a favourable energy balance. The main disadvantages are that biotechnologies cannot be used for all pollutants, are still empirical processes, and are poorly understood by traditional control engineers.

Waste water treatment is by far the most developed sector of environmental clean-up. In this market, biotechnology already dominates all other technologies. The competitive advantages of biotechnological treatment include low costs, the fact that both very high and very low concentrations of pollutants can be handled, and the production of biogas where appropriate. The disadvantages include sludge production and disposal, lack of operational stability and predictability, and the fact that not all pollutants can be removed.

In biological soil treatment, industry offers mainly services, rather than systems or installations. Compared to other technologies (incineration, landfill), the applications of biotechnology are still limited, but are growing rapidly. Advantages of biological *in situ* cleaning include low costs (again, approximately 60-80 per cent lower than physical and chemical methods) minimal site disruption, and complete degradation of pollutants into harmless substances. The drawbacks include the fact that complex mixtures of contaminants are difficult to treat, that laboratory conditions are not easily translated into large-scale practice, and that the selection of microbial strains is based on trial and error. Also, processes are slow and do not reduce pollutant levels to zero.

Industrial biotreatment of solid wastes is largely limited to the composting of wastes with a high proportion of organic materials. Future applications will include selective concentration of pollutants and conversion of highly organic solids into biogas. The competitive advantages include moderate costs, significant reduction of overall waste volume, and the possibility of selectively concentrating and removing pollutants; the disadvantages are slow reaction rates, and the fact that some wastes are not easily biodegradable.

Thus, the industrial application of environmental biotechnology varies substantially among sectors. However, industry expects an increasing role for biotechnology in all sectors. Biotechnology will not be the only solution, but it will be an important tool within a broader set of technologies. Key to further development is better integration of several biotechnological sub-disciplines.

Following this review of products and services, Chapter 8 contains an extensive analysis of market size, supplier companies and competition factors for each of the four treatment sectors. The environment industry has a dual structure, with a few large companies accounting for approximately half of total output, and a large number of very small companies accounting for the other half. The diversity of the environment market, and the relatively low barriers to entry, explain the large number of supplier companies. Their numbers are estimated at approximately 30 000 in the United States and Canada, 20 000 in Western Europe and 9 000 in Japan. The size of the total environmental market of the OECD area was estimated at $200 billion in 1990, growing to $300 billion by the year 2000, which means that the environment industry is already an important factor in GDP, employment, and trade. Overall, an estimated 15 to 25 per cent of the total current market, or approximately $40 billion, is biotechnology or contains a biotechnology component.

In the air/off-gas sector, markets are small but growing fast. In water treatment, markets are huge, with investments and operational expenditures in the OECD area reaching tens of billions of dollars annually, although saturation will now lead to slower growth. In soil treatment, actual markets in the OECD area are expected to increase from approximately $10 billion in 1990 to $20-30 billion in the year 2000, but potential markets could be enormous (several hundred billion dollars), considering that several hundred thousand sites should be cleaned up. In the solid waste sector, markets are expected to increase from $35 billion in 1990, to somewhere between $50 billion and $160 billion in 2000.

On the supply side, the air/off-gas sector has relatively few competitors (approximately 500 in Europe), but competition is intense and based on ability to comply with government standards and on price and reliability. The water treatment sector attracts large numbers of firms; competition is severe (saturation) and largely based on price, although know-how, image, and good links with authorities and users also play a role.

In the soil treatment sector, many small firms can still be found, but large chemical and energy companies are now entering the field and will absorb the small ones. In the end, only a few large groups provide a broad range of services and guarantee compliance with regulations will remain. Competition is severe and based on price, and considerable over capacity is building up. In the solid waste sector, the same trend towards industrial concentration can be observed as regulations become tighter. Competition is moderate (except for suppliers of composting units), and based on price, service, and compliance with regulations.

Chapter 9 describes the five distinct types of supplier companies:

   i) Large multinationals, in the chemical, pharmaceutical and agro-food sectors, where environmental biotechnology does not yet belong to the core business and bioremediation R&D is small (between 1 and 2 per cent of total R&D). However, these companies spend more and more on developing "clean" technologies.
   ii) Small, R&D-based companies spending up to 75 per cent of annual turnover on environmental biotechnology R&D, and focusing on niche markets.
   iii) Engineering, equipment and construction companies that manufacture or install process equipment. These are usually large but spend little in house on environmental biotechnology R&D (less than one percent of total R&D). They typically organise co-operative R&D projects with universities and public R&D institutes.
   iv) Large building, earth-moving and infrastructure companies with special soil remediation departments. Their characteristics are very similar to type *iii)* companies.
   v) Consultancy firms offering advice, but rarely carrying out R&D.

Environmental technology companies have a tradition of close vertical co-operation and integration, linking suppliers, engineering and manufacturing companies, as well as end customers. Integration expresses itself in numerous joint ventures, and co-operation across national borders is intense. "Globalisation" is one of the main characteristics of environmental biotechnology.

In spite of obvious advantages of environmental biotechnology, its diffusion through industry is not yet fast. When biotechnology competes with other technologies, the decision of the industrial manager often depends on his educational background, which all too often does not include biological or ecological sciences. Thus, biological pollution treatment is sometimes considered more difficult simply because the polluting company has no biotechnological expertise.

Chapter 10 presents the environment in which these companies operate, as well as their views on government policies and support. None of the companies contacted (only European and American ones) considered national or international government support as indispensable for their own R&D or demonstration activities. The most important positive factors for the development of environmental biotechnology were said by these companies to be legislation and enforcement. This is followed by support for scientific innovation and commercial implementation. The most important negative factors are absence or unpredictability of environmental legislation, followed by economic factors, such as insufficient return on investment, foreign competition and market saturation.

Clearly, it is governments at national and international levels that presently determine innovation in environmental biotechnology, by directing and shaping the demand for biotechnology products. Also, they decisively affect the economics of pollution control. The main problems here stem from lack of co-operation and inconsistencies between government departments.

In many countries, legislation is creating the new market opportunities for environmental biotechnology and will remain a decisive factor in the future. Financial support for academic R&D, innovation, and demonstration will become more important with the maturing of the environmental biotechnology sector.

Industry would like to see more *consistent* public sector R&D support, both for long-term fundamental science and short-term demonstration projects. A number of recommendations were made by the companies contacted, which requested the provision by governments of:

   − international priority setting and action plans;
   − consistent and harmonised legislation;
   − proper enforcement;

- no dramatic and unexpected changes in legislation;
- support for implementation of environmental biotechnology.

In addition, there is industrial support for the development of process-integrated technologies, in addition to add-on clean-up technologies and for public sector support of education and the raising of public awareness.

Part I

THE STATE OF THE ART

Chapter 1

Background

1. Traditional methods of waste treatment

Human domestic wastes, both liquid and solid, have been, and in many parts of the world still are, discharged into rivers and the sea, spread on land, or deposited in landfills. Water is indispensable for all forms of life, yet fresh water resources, particularly, have been grossly abused due to industrialisation and resultant changes in population distribution. Only about 10 per cent of the global production of waste water is subjected to some degree of treatment, while the remaining 90 per cent places excessive demands on natural self-purification capacities. While these self-purification mechanisms have allowed widespread reutilisation of water after passage through the natural environment, human intervention has reduced the residence time of water in hydrological cycles.

We rely heavily on biological systems to break down waste materials and have come to assume that virtually any material discharged into the environment will eventually disappear. To some extent, this is true. Plants and microscopic organisms have a vast capacity for rapidly degrading organic materials. Biological waste treatment, protection of drinking water sources, and disinfection of drinking water and sewage were gradually introduced in the early part of this century.

Traditional waste management operations include treatment of domestic and industrial sewage in sewage plants, landfilling and incineration of solids, and the disposal of sludges and animal slurries on the land; in all, with the exception of incineration, biological processes play a role and may be regarded as extensions of the natural activities of the biosphere.

In developed countries, most urban wastes are released into sewers, collected and disposed of as solid or semi-solid garbage or discharged into the air. Untreated sewage is still discharged directly into the sea, although declining. Domestic sewage is reasonably constant in composition and contains few really toxic materials, but the presence of variable industrial wastes may seriously complicate the treatment process.

The treatment of domestic sewage has several objectives:
- the removal of readily oxidisable organic materials which contribute to biochemical oxygen demand (BOD);
- the removal of solids in suspension which would otherwise result in the silting of waterways;
- the removal of toxic materials, such as ammonia, arising from the biological degradation of natural wastes;
- the removal of heavy metals and other industry-derived pollutants;
- the elimination of pathogenic micro-organisms that might infect others downstream.

In the preliminary and primary stages of treatment, traditional waste water treatment technology removes suspended solids by a sequence of mechanical and physical processes. Secondary stages are biological in nature while, today, because of the recalcitrant nature of many pollutants present in both sewage and industrial waste waters, physico-chemical, anaerobic biological, and other tertiary stages are finding increasing application. It is common to use only primary physical treatment where the effluent is to be discharged into the sea. Organic solids derived from secondary stages may be settled out and these, together with bacterial remains, are further digested to a final harmless material which can be landfilled, dumped at sea, used as fertiliser or even recycled as animal feed if the heavy metal content is acceptable. Slurries from animal husbandry are, wherever possible, returned to the land. Too often, however, excess slurries find their way into rivers. An alternative treatment, used also for digestion of municipal sludges, is anaerobic digestion, which in one step can lower total solids, reduce odours, kill pathogens, and produce methane. The methane has energy value and may be used as a fuel; this is one of the major attractions of anaerobic processes.

Household solid refuse contains largely human, food and packaging wastes, much of it biological in origin and hence biodegradable. The solid and semi-solid animal and plant remains, together with paper and paper products in domestic waste, have customarily been deposited in landfills. While some of this material is degraded anaerobically (demonstrated by the formation of methane), there is plenty of evidence that organic waste can also be preserved in landfills. Excavation of such fills has revealed perfectly readable fifty-year-old newspapers.

The airborne pollution, formed as gaseous products of waste decomposition, has traditionally been vented to atmosphere or led into a combustion process – usually the plant boiler. Rather more recently, extensive beds of materials such as peat and heather have been used to absorb odours (see Chapter 2).

Industrial wastes are more site-specific in composition than their domestic counterparts. Many chemical wastes have been disposed of in landfills or incinerated. Much industrial waste has in the past been dumped, often illegally, in pits and elsewhere: the rates of breakdown are likely to be low, but seepage into the surrounding soil may take place. The food, natural fibre, wood and paper industries, which use natural raw materials, generate biodegradable processing wastes; however, some manufacturing involves preservative chemicals which may inhibit subsequent biological breakdown.

2. Nature of the problem

Wastes

In terms of quantity, the most common wastes are those derived from agriculture, domestic refuse and sewage, and industrial effluent (Table 1). These contain large quantities of organic matter, which, if released into the environment untreated, would cause extensive pollution. The biological oxidation of this organic matter by the natural micro-organisms in the environment causes rapid oxygen depletion and consequent death of higher organisms. Agricultural run-off in all forms is one of the most intractable forms of pollution because of its extended nature. The slurry from farming and intensive animal rearing is spread on the land, and rain and irrigation carries the excess, as well as pesticides and fertilisers, into water courses. Mining activities, and the natural effects of micro-organisms on metal ores, may also give rise to polluting run-off. For more details on agro-food biotechnology and biomining, see Appendix 1.

The development of environmental control techniques, sewage treatment for water-borne wastes and landfill for solids has gone some of the way towards solving the problems associated with waste disposal but has not eliminated them. Indeed, they have given rise to secondary sources of pollution. Sludge from sewage plants concentrates metals from industrial waste to the point where it may no longer be spread safely on land; landfill generates methane and seepages of toxic leachate which return to rivers and groundwater.

The adequate treatment of sewage to remove pathogens, reduce nutrient levels, and eliminate toxic compounds and the disinfection of drinking water have become very important factors in lowering rates of waterborne infectious diseases. The installation of advanced sewage treatment facilities permits the reduction of the biochemical oxygen demand (BOD) of sewage effluents and is essential for the abatement of eutrophication problems in aquatic ecosystems.

Table 1. Annual waste generation
(OECD countries, late 1980s)

Type	Million tonnes
Municipal waste	420
Industrial waste	1 430
Hazardous and special waste	303
Air emissions (excl. CO_2)	214
Carbon dioxide	6 256
Oil pollution	3.5
N fertiliser (applied)	70
Lead	2
Pesticides	0.075

Source: M. Griffiths, based on OECD data.

Since biological treatments have long been used for waste disposal, what new problems force the consideration of new biotechnological solutions? First, there is a need to supplement the natural biodegradation processes that occur when dilute wastes are released into the environment. Managed biological systems are required when higher concentrations of wastes are destined for disposal into receiving waters or onto land with insufficient degradative capacities. The larger amounts of waste associated with greater human populations in the growing urban centres require more efficient treatment systems and new waste treatment reactor designs, and these in turn need new biotechnological principles. In consequence of the public demand for clean water and land, the results of waste treatment have to be superior to those of the traditional reactors in use today. Multiple uses of water and land and the decreasing availability of disposal sites make waste minimisation and the efficient disposal of wastes a necessity. In general, new management systems, including novel bioreactor designs and increased reliance on composting, are likely to be required for the general reduction of domestically generated wastes.

In addition, the materials used in modern society present special disposal problems. Plastics and other materials routinely used in households escape degradation in most waste treatment facilities and build up in locations where they are discharged. Landfills accumulate many resistant wastes rather than act as sites for their degradation. Industrial waste streams also present challenges to traditional waste treatment facilities, forcing in many cases separate treatment and disposal options. Typically, the focus on industrial wastes concerns specific compounds that may be toxic or otherwise create unacceptable risks to human health and the environment. Novel and segregated treatments employing specific biotechnological disposal options may be required to treat each waste.

Pathogenic organisms

Any untreated waste water stream, domestic or industrial, is likely to contain organisms pathogenic to man and animals. A major objective of treatment processes should be their removal. These organisms include bacteria, fungi, protozoa, worm eggs, and viruses. In primary sedimentation, about half the pathogens are removed in sludge which is treated and stabilised. The effluent of this stabilisation contains about the same concentration of pathogens as the raw waste water. Pathogens remain active unless heat-treated. Anaerobic treatment of sludge is regarded as having a much slower kill rate than aerobic treatments such as composting.

Secondary sludge has a much smaller volume and many fewer active pathogens, but the treated waste water still is not free of pathogens and requires a tertiary treatment. This may include flocculation, precipitation and filtering through sand filters. Disinfection of the treated water with chlorine or an equivalent is also practised. It is generally assumed that pathogens are either absent or present in very low concentrations in aerosols produced by aeration, and action is rarely taken to eliminate this aerosol formation. The concentration of pathogens in waste sewage sludge suggests that it must be handled judiciously. By far the greatest scope for treated waste water reuse exists in irrigated agriculture; the dangers of reutilising waste water include the transmission of infectious diseases by pathogens that become associated with the products of irrigated agriculture and aquaculture.

There are few reliable data on the infective doses of the traditional bacterial and protozoan pathogens. Viability can be determined, but what is needed is a measure of the virulence of the various strains and this cannot be determined *in vitro*. Methods for detecting of pathogens are available but many parasites and viruses have a much higher stability than respiratory and enteric bacteria. There is therefore a need to develop test procedures for the rapid identification of all types of potential pathogens, including newly evolved strains. Regulations may be necessary not only for the kind and quantity of pathogens in polluted soil and water but also for the microbial composition of the cleaned matrix.

Pollutants

Irrespective of where and when it occurred, industrialisation has been closely linked with environmental pollution. Industrial pollution is multi-faceted: Consumption of fossil fuels due to transport yields organic vapours, oxides of nitrogen and sulphur, ozone, and carbon dioxide; the fuels themselves, if spilled, can be environmentally disastrous. Industrial production wastes include acidic gases, toxic organic vapours, solvents, chlorinated organic molecules, and metal salts in waste streams. Losses and residues have always been part of industrial production. Typically, 20 per cent of raw materials may be wasted. Many of the losses can be attributed to unused raw materials and products which can be easily recycled, but about 7 per cent occur in dilute form from plant cleaning, spillages, and escapes. Leakages from plant can give rise to contaminated soil, and polluting substances can percolate through the soil and reach groundwater and hence aquifers. Inadequately controlled dumping of industrial solid waste can permit the spread of toxic materials into the wider environment.

The changing nature of pollutants from "natural" to xenobiotic chemicals may exhaust self-purification capacities. A significant number of synthetic compounds, particularly those that are not related to natural compounds, persist in the environment, as modern sensitive chemical analysis reveals. There are perhaps 100 000 man-made chemicals in commercial use, and hundreds of new ones are produced each year. They represent a massive influx of xenobiotics into the environment. Major areas of concern are pesticides, polyaromatic and polychlorinated hydrocarbons and organic solvents. Many dyes and colours and some modern detergents are also only poorly degraded by conventional waste treatment processes, while many synthetic products and intermediates of degradation are not detected by the usual analytical methods.

In addition to organic materials, inorganic pollutants, such as the plant nutrients nitrogen and phosphorus and also metal ions, constitute an environmental problem. Domestic wastes contain large quantities of nitrogen as urea and protein and these are rapidly metabolised to ammonia. Ammonia is toxic to aquatic life above 2 mg/l, and residual ammonia in effluents and landfill leachates has to be removed by oxidation to nitrate.

Nitrate itself accumulates in natural waters because of the run-off or leaching of a significant fraction of nitrogenous fertilisers and organic wastes applied to arable land, and potential health problems are said to be associated with this material, although it should be noted that this is controversial. First, nitrate may interfere with the oxygen-carrying capacity of the blood in infants (infantile cyanosis, methaemoglobinaemia or blue baby syndrome). Second, it is claimed that nitrate can, under certain circumstances, be reduced to potentially mutagenic nitrosamines in the body. The World Health Organisation (WHO) has recommended a limit of 50 mg/l nitrate in drinking water. This standard is incorporated into the European Community drinking water directive but the EC guide value is even lower, at 25 mg/l. The scientific basis for these figures is strongly disputed.

Phosphate, like nitrate, is a residue from the biological oxidation of organic matter but is also derived from domestic and other detergents. It may be present in settled sewage at concentrations that markedly exceed those that can be removed by typical bacterial growth processes. Although phosphate levels in sewage can be significantly reduced by substituting other chelating agents for polyphosphates in detergents, levels may still exeed what can be removed by typical bacterial growth.

In combination with nitrate, phosphate can lead to nutrient enrichment and excessive algal growth (eutrophication). These algae are unsightly but may also cause problems of taste, odour, oxygen depletion and occasionally toxins in water to be reused for domestic supply. Phosphorus is often the limiting plant nutrient for the growth of phytoplankton (algae and aquatic plants); for example, 1 mg of phosphorus can generate 100 mg of algae (as dry solids) and eventually, on degradation, 140 mg of oxygen demand.

In most industrial cities and towns, effluents may be contaminated with toxic metal ions. The origin of much heavy metal pollution is trade effluents, particularly from the metal finishing industries, that are discharged into the sewers. The metal concentrations of such effluents are usually too low to permit economic and/or effective recovery by conventional technologies. The main concern is that if metals are not eliminated during treatment, they will be discharged into natural waters that are used as sources for drinking water. Metals such as zinc, copper and nickel, for example, are phytotoxic, their effects are additive and the risk arises from accumulation in the food chain. Cadmium, lead and chromium are more mobile, and consumption of contaminated food presents a direct risk. There are other potential risks from mercury, molybdenum, arsenic, fluorine and selenium, among others. Most of these elements can also be present due to aerial deposition as well as applications of surplus sludges to soils and leachate contamination of water. A more recent problem, associated with the use of nuclear processes both as a source of energy and in medical facilities, is the presence of radionuclides in industrial effluents.

Cyanide, which plays an essential part in a number of industrial processes, particularly in the metals industries, may be found in effluents. Concentrations as high as 200-500 mg/l have been measured, limit value stipulations are 0.1-0.5 mg/l. Extreme cyanide concentrations can also be found at sites that were once gasworks.

Land contamination

Contamination of land has been a consequence of industrial activity since the start of the industrial revolution. Despite increasing precautions, spillages occur, and oil and chemicals seep into the soil beneath the plant. Such sites, some identified, many more not, exist in all industrialised countries. We have in the past relied on natural organisms to degrade these soil pollutants, and such sites are, in fact, an excellent starting point for the search for degrading organisms. In the past, polluted soil has been excavated and dumped elsewhere – in landfills or just into holes in the ground. Degradation *in situ* has been stimulated by the process of landfarming, in which the soil is ploughed repeatedly to aid aeration and encourage the indigenous microbial population to break down the organic molecules. In many countries, wastes from this process are still considered to be hazardous and carry the risk of toxic chemicals being leached into aquifers and other sensitive environmental structures. Landfarming is therefore a process which has to be carefully monitored.

Pollutant differentiation

In most industrialised countries, legislation targeted at environmental protection has required both increased efficiency and wider application of waste water treatment processes. Pollutant loads in waste water have traditionally been specified in terms of parameters such as BOD or chemical oxygen demand (COD), a meathod which precludes any differentiation among the various polluting compounds present. While such differentiation is usually thought to be unrealistic in municipal sewage, it is an essentially realisable objective for industrial waste waters such as those from the petrochemical and other narrow product spectrum industries. Further, in the case of the noxious priority pollutants, differentiation is essential if measures for effective environmental protection are to be introduced. Unfortunately, the recent heavy emphasis on priority pollutants has also resulted in a marked neglect and subsequent ignorance of essential details concerning the biodegradation of many extensively used, relatively less noxious, bulk chemicals.

Suitability for treatment

In many industrial treatment plants, organisms are subject to variable operating conditions, particularly with respect to influent flow rates and pollutant concentrations. The waste water, as received at the treatment plant, is often entirely unsuitable for effective biotreatment. For optimum performance, process cultures of the type used in industrial activated sludge plants require the process influent to be within the limited temperature and pH ranges suitable for organism growth, with ratios of pollutants to nutrients that are also appropriate for growth.

Analysis of many industrial waste waters indicates that, although the bulk load of carbonaceous pollutants present is potentially biodegradable under ideal conditions, their temperature is frequently too high, their pH is often in the range where only acid-loving micro-organisms thrive, and there are deficiencies with respect to essential nutrients, so that, prior to biotreatment, many properties need adjustment. In addition, many carbonaceous chemicals inhibit microbial growth, while others are recalcitrant – are not easily biodegraded – although they may, like heavy metals, be accumulated by microbial flocs, thereby becoming a sludge component.

3.   Science and engineering

Biological reactions

In lay terms, fermentation, is any industrial biological process that takes place in a vessel or reactor, *e.g.* for the production of antibiotics – antibiotic fermentation. Fermentors are the vessels in which fermentation takes place. However, the scientific definition of fermentation is somewhat different: an anaerobic metabolic process in which the carbon food source is also the terminal electron acceptor. It is thus the opposite of respiration, which requires an external electron acceptor, and can be aerobic, using oxygen, or anaerobic, using a variety of inorganic materials. Alcohol production using yeasts, which is an anaerobic process, is true fermentation. In order to avoid confusion, the term fermentation will not be used here and all contained reaction vessels will be referred to as bioreactors. In the field of bioremediation, the definition of a bioreactor has often been widened to include piles or heaps of material, landfarming and even *in situ* processing. In this report, these are not included.

In order to carry out industrial biological reactions in a reproducible fashion, it is necessary to be able to control the environment in which cells convert substrate to product. In practice, therefore, biotechnological processes are normally carried out in a contained system. In general, the biological materials employed have to be maintained in a very constant and/or controlled environment: Parameters such as temperature, pH, and oxygen concentration may be closely monitored and controlled throughout the process. Processes performed with single cultures normally have to be carried out under strict (aseptic) conditions in order to maintain the integrity of the reaction.

Applications of microbes in industrial processes include the production of low-cost high tonnage (bulk) chemicals such as ethanol and methane, and also medium-cost bulk products used in food and medicine, such as citric, gluconic and itaconic acids. High-value products such as amino acids, vitamins, steroids and antibiotics are also produced by microbial reactions.

The commercial exploitation of isolated microbial enzymes dates back to 1894, when Jokichi Takamine patented a method for preparing diastases (a particular group of enzymes) from moulds. Since that time, enzymes have been separated from plant and animal tissues and from selected micro-organisms and subsequently used to carry out a multitude of biochemical transformations. Enzymes are used in a large number of industrial sectors, of which the largest is the food industry. Other areas include detergents, medical pharmaceutical products, and textiles. In many instances, enzymes are favoured over traditional catalysts in view of their high catalytic power, broad range of activity, and mild operating conditions (including pH, temperature, and pressure). Enzymes may eventually be important for primary attack on pollutants, leading to compounds which are more easily degraded.

Use of natural organisms

The first step normally taken when developing a biotechnological process is to identify a suitable organism. Once it has been found, a variety of methods are used to produce greater quantities of the desired product. Techniques for procuring new micro-organisms from the environment include: choice of habitat for sampling (possibly in a polluted area); physical separation procedures for isolating the desired micro-organism, and choice of method employed to achieve selection (typically using enrichment cultures – media in which some organisms have an advantage over others). The random screening of isolates can be very time-consuming, labour-intensive, and costly. Consequently, the choice of source location, selective pretreatments, media, and incubation conditions are of paramount importance to success. Once potentially valuable strains have been selected, the cell genome may be improved in a number of ways, by conventional breeding or by genetic modification. The resultant improvements may apply not only to the selected reaction but also to stability in growth and increased resistance to by-products.

A fundamental problem for assessing the total microbial resources available for biotreatment processes is the inability to isolate or characterise most of the microbes present in a given habitat. Traditional techniques for analysing the composition of complex mixed microbial populations require that any organism present must grow and multiply under laboratory culture conditions. Further, a prerequisite of traditional characterisation procedures is that similar strains must previously have been characterised. Estimates of the undiscovered microbial resources in natural ecosystems suggest that they probably represent a high percentage of the total microbes present in such ecosystems. Recently, molecular genetic techniques for exploring the extent of undiscovered microbial resources have become available. Using one such technique, it was shown that in a microbial community that had previously been well characterised using conventional methodology, more than 80 per cent of the microbes present had yet to be isolated and characterised.

Application of micro-organisms in waste treatment

Most industrial biological reactions used to produce of useful products involve single species (monocultures). In other instances, the action of a single species may predominate although a mixed microbial population is actually present. In nature, mixed microbial populations are ubiquitous in air, soil, and water. Microbial consortia also grow on and inside higher organisms. Reactions which take place in vessels or locations open to the air, *i.e.* most waste treatments, and those occurring in water or soil are invariably interactions of many single and multi-cellular organisms, and from both the plant and animal kingdoms. It is important to note that applications of bioreactors for the remediation of waste materials frequently depend on the use of a variety of mixed cultures.

Traditional biological treatment processes for wastes rely on the activities of many species of organism within a complex community. Numerous bacterial, fungal, and other microbial populations may simultaneously act upon classes of compounds of proteins and carbohydrates to bring about their decomposition. The complexity of these communities, in the biofilms of trickling filters and the even more complex activated sludge suspended communities, for example, has made it difficult to understand the roles of individual microbial populations. In essence, the community has been regarded as a superorganism, and, as long as the overall amount of organic matter diminishes at acceptable rates, it is seen as functioning acceptably. When the system fails to achieve the expected endpoints, the entire community is replaced; for example when an activated sludge facility goes sour and fails to reduce the biochemical oxygen demand, the sludge is replaced so that a new microbial community is put to work.

As the focus of waste treatment has been extended to specific pollutants, it has become necessary to understand the physiological interplay within microbial communities and to investigate the abilities of populations to degrade specific pollutants. It has become important to examine the roles of individual species and the joint activities of a limited number of species acting in concert within consortia in order to design better bioreactors and to permit the treatment of pollutant-containing wastes and polluted water, soil, and air. In this context, it is the metabolic activities and the underlying genetics controlling the capacity and expression of degradative activities, rather than specific species, that are important in process development and bioreactor design. The role of microbial consortia is further discussed in Appendix 2.

In some cases, if the physiologies of communities involved in the degradative processes are understood, rates of decomposition can be optimised. This is certainly the case for petroleum hydrocarbon pollutants, where nutrient addition and oxygenation have been successfully employed to stimulate the degradative process. Provision of both aerobic and anoxic zones within sewage treatment facilities has permitted the combined reduction of organic matter and the removal of nitrogenous inorganic compounds. Understanding that denitrification was an anaerobic process made it possible to alter conventional sewage treatment reactors in order to allow this process to occur.

Since xenobiotic compounds do not occur naturally, it is not surprising that micro-organisms have not evolved extensive capabilities for degrading these materials and that few populations within natural communities can do so. Thus, processes for degradation of specific xenobiotics, such as polychlorinated biphenyls (PCBs), trichloroethylene (TCE), dioxins, and the like, rely upon carefully selected micro-organisms or consortia. Maintaining such populations requires careful management in order to permit the persistence and expression of the desired gene activities.

Recombinant DNA technology

Just as chemists have circumvented natural product formation in producing xenobiotics, molecular biologists are able to use recombinant DNA technology to form organisms with novel degradative capacities. Genetic recombination involves a restructuring of DNA molecules so that new genomes containing information from different DNA sources are formed. A variety of mechanisms permit the transfer of genetic information from a donor to a recipient micro-organism. In bacteria, DNA may be transferred by conjugation, which involves direct contact between the donor and recipient strains, transduction, which utilises a virus as a vector for carrying the DNA from the donor to the recipient strain, and transformation, which involves the uptake of naked DNA by a competent recipient strain. Micro-organisms, when produced in this way, are referred to as genetically modified organisms (GMOs) or, in the United States, as genetically engineered micro-organisms (GEMs). It should be noted that all of these processes go on in nature, albeit in a random way. Often the key to engineering an organism for biotechnological application has to do with gene expression and allowing the engineered organism to survive and express the activities for which is has been designed.

Genetic exchanges between distantly related organisms are known to occur in nature. Recent evidence suggests that widely different species of bacteria easily exchange genetic material in the environment, and even wider transfers (e.g. between bacteria and yeasts) have been demonstrated to occur under optimal laboratory conditions. One particular genus of bacteria, *Agrobacterium*, is noteworthy for its ability to infect and to transfer genetic material into the cells of dicotyledonous plants. However, there is no evidence at this time in the scientific literature that other exchanges of this type, or non-sexual exchanges of genetic material between higher organisms, occur frequently in nature.

When considering the potential applications of GMOs versus naturally occurring ones, questions of cost (e.g. R&D and also regulatory costs), novelty, and specificity arise. In many cases organisms have already evolved the necessary capabilities, and it is the engineering that provides the basis for biotechnological applications. The environmental clean up of fuel-based hydrocarbons is a good example of the application of naturally occurring organisms for environmental clean up where recombinant organisms may not be needed. In other cases, naturally occurring organisms with the desired capabilities have not yet been identified, and it is here that recombinant DNA (r-DNA) technology may contribute to the environmental applications of biotechnology. This may involve the formation of highly active or overproducing strains that can alter the economics of using biological as opposed to physical (chemical) methods, the formation of strains with novel activities, or of strains that carry out very specific transformations to produce environmentally acceptable materials. The degradation of chlorinated hydrocarbons and the production of biopolymers are areas in which genetic modification through r-DNA technology may be especially useful.

There is debate over the need to use r-DNA technology to develop micro-organisms capable of degrading the more recalcitrant pollutants. Given sufficient time, evolution may well lead to the formation of similar organisms that can transform xenobiotic substances; r-DNA technology can, however, achieve greater speed and efficiency. The process of "*in vivo* engineering", in which organisms are selected by applying a specific environmental stress, may yield the desired results without any problems of public acceptance. *In vivo* transfer, which may occur when genetically diverse micro-organisms coexist, has been used to obtain microbes that degrade previously recalcitrant compounds. Dioxin degradation may be an example of early public acceptance of the use of GMOs. This subject is somewhat controversial because genetic engineering does not inevitably lead to organisms that duplicate naturally occurring processes. It should be noted, however, that for certain uses of GMOs, for example in the pharmaceutical industry, controversy has decreased as new and useful products have been developed and marketed.

An important, recently developed, technique in the field of genetic modification is the use of so-called suicide genes, whereby a reaction controlled by the gene can either stop essential growth processes or even cause disruption of the cell following release. This technique is still at an early stage of laboratory experimentation but holds promise for the control of released organisms.

The introduction in the early 1970s of new molecular technologies to produce organisms with novel genetic combinations initiated discussions on safety in biotechnology. The Asilomar conference in 1975 was the basis of the NIH guidelines describing the conditions under which rDNA experiments – at that time limited to contained sites – should proceed. These US guidelines were widely followed on a world-wide and voluntary scale. They have been progressively relaxed in view of the growing safe experience with rDNA technology.

In the 1980s, the OECD report *Recombinant DNA Safety Considerations* (the so-called "Blue Book") was one of the first international scientific frameworks for the safe use of organisms derived from r-DNA techniques in industry, agriculture and the environment. Due to a lack of sufficient experience, and out of a desire for prudence, the OECD recommended at that time both a rigorous progressive and case-by-case approach. The first principle involved the application of a gradual and successive scaling up (*e.g.* laboratory, greenhouse, small scale plot, commercialisation). The second principle took into account the particularities of each case, that is to say, the combination of organism, trait, application, and environment and the need for their proper assessment as well as their interactions.

The OECD Group of National Experts on Safety in Biotechnology has worked since April 1988 to update and develop further the principles set out in 1986. In the meanwhile, r-DNA techniques were considered increasingly to represent an extension of conventional genetic procedures, and r-DNA organisms to present risks that are the same in kind as those associated with other organisms, and controlled by the same physical and biological laws.

Moreover, in the light of the growing experience and know-how with certain organisms, traits, applications and environments – as well as possible combinations thereof – the OECD Group accepted in 1992 a concept of "familiarity" that permits a certain rational flexibility with respect to the two existing OECD approaches. Thus, the number and modalities of the stages with progressive scale-up is not strictly fixed but can be modified in light of the degree of familiarity with the category of cases (mainly combinations of organism, trait, application, and environment) to be dealt with from the point of view of risk management.

Bioengineering

Environmental biotechnology, like all other technologies, is a composite of disciplines; it requires inputs from soil geochemistry, on the one hand, to fluid mechanics, on the other. The discipline of chemical engineering, is a major contributor to understanding processes; it ranges from industrial chemistry, at one extreme, to physical process engineering, at the other. As far as biochemical (bioprocess) engineering is concerned, a major problem is a clear definition of what actually constitutes the activity, on the one hand, and whether or not this differs from more established civil, sanitary and environmental engineering, on the other.

The pragmatic nature of traditional waste treatment has meant that civil engineering has had a significant influence on plant design. Increases in the quantity of waste to be treated have often resulted in increased plant size rather than increased productivity. Closer co-operation between engineers and biotechnologists is now leading to smaller, modular plants, a process which will continue as we gain a better understanding of the underlying biology.

In designing a novel reactor, there is a tendency to regard micro-organisms as "bags of enzymes" that do not change in composition and activity and, as far as engineered processes are concerned, are of constant size and morphology, irrespective of their growth rate. In practice, this is not the case. Modern biodegradation research has concentrated on identifying detailed metabolic pathways rather than obtaining kinetic information, in spite of the fact that the efficiency of microbial processes depends on the rates at which each of the enzymes in these pathways act. Thus, all the enzymes forming a specific pathway are coupled to one another by the metabolites they share, and the rates and fluxes through particular enzymes are significantly affected by adjacent enzymes. When considering the biodegradation of an individual pollutant, it is essential to examine the system as a whole rather than the individual enzymes and the specific reactions they mediate in isolation.

Three of the least appreciated concepts in microbial physiology are: no microbial system is ever in a steady state but changes with respect to time; gradients exist in all microbial systems which create conditions at reaction sites that are totally different from the conditions that can be measured either by probes inserted into or samples removed from systems; and when the physiological states of the same strain differ, their dynamic response to identical perturbations will also differ.

Chapter 2

Emerging Biotechnologies

1. Prevention

A new philosophy is emerging in the field of industrial production, that of keeping a balance between development and environmental preservation. It is replacing the traditional approach of using up resources as a consequence of ever higher process throughput. This philosophy reflects the increasingly popular theme of sustainable development, and waste minimisation and recycling efforts are now widespread. While the search is on for clean technologies which produce no waste, this is recognised to be a long-term goal. The development of any new process is expensive, and new biological processes may only be considered when the conventional equivalent is nearing the end of its economic life. Until that time waste treatment is usually the less costly option. However, the concept of economic life is determined not only by physical wearing out but by comparative costs, and thus a sufficiently cost-effective alternative may curtail the life of an existing process.

Industry, taking a more immediate view, defines pollution prevention in terms of any technology that restricts the release of pollutants into the environment. By this definition, pollutants which are contained within the manufacturing process and treated, recycled, or converted prior to their release to the environment are included. The incentives for bringing treatment systems inside the factory fence may be economic or may stem from public pressure, and the recent involvement of major companies in the voluntary reduction of key pollutants has created additional widespread interest in waste minimisation technologies. The benefit to the corporate image of advertising the use of these technologies should not be neglected.

Legislation plays a significant role both in pollution prevention and recycling. For the latter, for example, there are now ordinances in some countries requiring used packaging materials to be taken back by suppliers. The role of the automobile industry, in which biodegradable parts are playing an increasing role, is noteworthy, particularly in Germany. Here also, the public relations effect is considerable.

Biotechnology can be used in a number of ways to prevent environmental damage: for example, a biotechnological process can be used to convert a waste stream into value added products such as methane (biogas). Enzymes can be incorporated into animal feeds; this may increase the availability of dietary minerals to the animal, and, at the same time, reduce the nutrient content of their waste. Processes using micro-organisms may also be used to recover valuable materials, particularly precious metals, from waste streams for recycling within the process. The simplest use is the "end of pipe" process, by which the waste stream is purified and the effluent released without harm into the environment. Process innovation, which is the development of new biological processes and the modification of existing ones, is an option, and the use of new biomaterials, which have little or no environmental impact, for example, biodegradable polymers, is another.

A good example of process innovation is the use of a biological reaction as a step in the manufacture of a major industrial chemical, acrylamide. Acrylamide is produced industrially as a monomer for synthetic fibres, flocculating agents, etc. The conventional synthetic process involves hydration of the nitrile with sulphuric acid and/or the use of inorganic catalysts. The inclusion of the enzyme nitrile hydratase in an alternative route is possibly the first use of biotechnology in the petrochemical industry.

In recent decades, an increasing number of biocatalysts have became available from screening (especially in Japan) and protein engineering. The production of enzymes has also become more efficient through r-DNA techniques. These developments improve the chances for more biotechnology-based production processes for a wide range of chemicals. In particular, the development of engineered enzymes may lead to spectacular process advances. However, new chemical catalysts are constantly being developed to compete with the biological ones, and the decision to use the biological system rather than a chemical alternative will be made on economic as well as environmental grounds. The role of biotechnology in cleaner technology is discussed more fully in Appendix 3.

2. Detection and monitoring

The purpose of detection and monitoring technologies is twofold: first, to provide prompt warning of pollution incidents and thereafter to assay the level of pollution generally or of specific pollutants; and second, where *in situ* remediation is carried out by specific micro-organisms, to monitor both the organisms and their effect on indigenous populations. These technologies are described further in Appendix 4.

Pollutant assay

The greatest threat to the producer of a waste stream is the unforeseen incident that results in a pollution event going undetected. Because acute pollution incidents are unpredictable, continuous or frequent sample monitoring to ensure detection. Water is already subjected to extensive biological analysis, using a wide range of tests including observations of genetic or structural damage in organisms, changes to microbial activity, and checking for alterations to species diversity.

Biological surveillance using field surveys of whole ecosystems (detecting changes in species diversity and numbers) has long been used to monitor chronic toxicity and bioaccumulation. Other examples of biological analysis are BOD and Biological Methane Potential (BMP). These tests are widely used by water authorities to assess treatability and toxicity of industrial effluents. BOD is also used as a measure of biodegradable organic compounds in waste water. Conventional evaluation takes several days and is thus unsuitable for process control, but a more rapid estimation is possible using microbial sensors containing immobilised whole cells on oxygen electrodes (Case Study 1).

Case Study 1:   Monitoring high volumes – United Kingdom

Monitoring BOD has always been difficult, as the standard method of determination takes five days. This is now seen as inadequate for many industrial processes, where effluent discharges vary considerably on a daily, if not hourly, basis. Rapid monitoring is essential for real-time pollution control.

A company in South Wales commissioned the construction of a modern, biological treatment plant to ensure improvement in the quality of the effluent discharged to the local river. However, the company decided that the conventional BOD test was a poor indicator of plant performance and investigated technology that would give instant measurements. The company opted for a German BOD analyser, supplied by another Welsh firm, which is an on-line monitor designed for applications where the effluent contains a high concentration of solids – an area in which monitors have failed in the past. Since the monitor was installed, the company has significantly reduced delays and costs associated with laboratory analyses.

The effectiveness of the plant was demonstrated when an unnoticed malfunction resulted in a very high BOD in the effluent. The organic load meant that the effluent treatment plant could not reduce the BOD to the consent limit of 5 mg/l. The monitor gave the company an immediate warning of the episode, which meant that the effluent could be recycled until the fault was rectified. Had the company had to rely on the old BOD measurement, there would have been a severe pollution incident and subsequent shut-down until standard tests confirmed that the discharge was within consent limits.

Source: Environment Business Newsletter, United Kingdom.

Other more specific analyses used routinely by the water industry include observation of chromosome damage to assess the risk arising from acute toxins. Biological activity is extensively used for monitoring water quality, and there are standard test procedures for assessing acute and chronic toxicity using a range of organisms such as *Daphnia* , algae, and even trout. These tests have also been developed to include monitoring the activity of caged fish and the breeding potential of mussels.

Two general classes of biological detection methodology, biosensors and immunoassays, are just beginning to be commercialised for a few chemicals, but for the most part, they are still in experimental development for environmental monitoring. In what seems to be a common path, these technologies were first developed for the health care industry but now are increasingly being adapted to detect environmental pollutants. Biosensors are devices in which a biological agent has been immobilised and incorporated as the sensing element. Established bioassay technology is linked with transducer technology to give rapid, easy to use, and often automated monitoring/analysis systems. The immunoassay is a powerful and versatile technique which has been successfully

Figure 1. Aldicarb RaPID Assay, an ELISA immunoassay which uses novel magnetic particles as solid support and means of separation

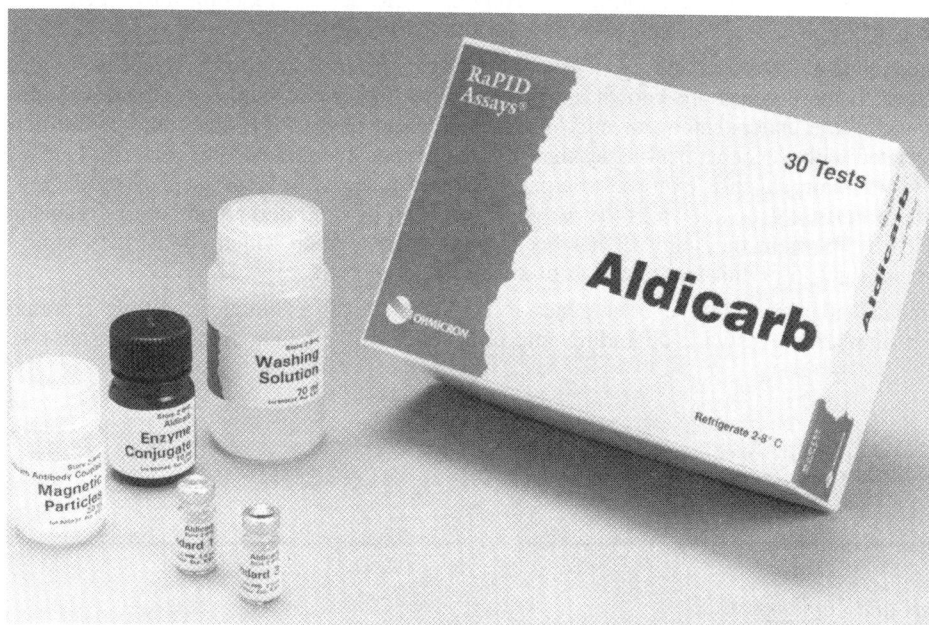

Source: J.T. Baker Inc., Europe; RaPID Assay is a trademark of Ohmicron Corporation.

used to measure a wide variety of compounds, primarily with medical applications. Immunoassays require highly specific antibodies albe to recognise and bind to single compounds, small groups of related compounds, or classes of compounds. The fundamental characteristics of each immunoassay are based on the specificity and binding affinity of these antibodies for the target compound (or compounds). Immunoassays have only recently been applied to the measurement of toxic compounds in the environment, and the potential for using immunoassays to solve some of the problems of environmental measurement is just beginning to be realised.

On-line monitoring is targeted at the detection of the "appearance" of pollutants/toxicants. This is a rapidly evolving area of biotechnology, and the sensitivity, accuracy, speed, and safety with which biosensors will be able to detect and measure biochemical analytes offer a wide range of potential applications, especially for the monitoring of groundwater and drinking water.

The determination of minute amounts of pesticides in water ($0.1\mu g/l$ of a single compound, $0.5\mu g/l$ of a mixture) required by the European Community Drinking Water Directive, for example, is a great challenge to the development of appropriate assays. Immunoassays, especially ELISA techniques (see Appendix 3), have received considerable attention in this context (Figure 1). The number of monoclonal and polyclonal antibodies raised against pollutants is growing continuously and these will be used in assays. They will also be candidates for biosensors once the technical problems associated with their immobilisation on suitable substrates is solved.

Specific biosensors, especially those based on individual enzymes, may detect single pollutants with exquisite sensitivity and are therefore complementary to physical techniques such as gas chromatography, which may identify a range of pollutants.

Micro-organism detection and tracing

In addition to the more traditional selection techniques and uses of special media, organisms may be detected and their numbers estimated by isolating their genetic material (DNA) and hybridising it with probe sequences which have been labelled with either radioactive atoms or with chemiluminescent dyes. A major use of this technology has been to isolate organisms with specific properties, for example naphthalene-degrading bacteria in activated sludge and pathogenic *Listeria* species in dairy products.

The development of the polymerase chain reaction (PCR) in 1983 was a major methodological breakthrough in molecular biology. PCR permits the *in vitro* replication of defined sequences of DNA whereby gene segments can be amplified. One application of this technique has been to enhance gene probe detection of specific gene sequences. In conjunction with DNA amplification using the PCR, detection kits for most *Legionella* species (the causative agents of Legionnaire's disease) are now on the market.

Using recombinant DNA technology, it is possible to insert into micro-organisms genes that can subsequently be detected; these genes are known as reporter genes or genetic markers. This development has been central in advancing the understanding of microbial genetics and physiology. The more useful genetic markers are those associated with a biochemical assay that is both inexpensive and easy to perform. One such is the *lac* gene which, when the organism is exposed to a particular substrate, will reveal itself by producing an insoluble blue dye. More recently, a gene known as the *lux* gene, has been inserted into a number of organisms which may then be detected by the light they emit. It has been claimed that certain genetic constructs using the *lux* gene produce sufficient light to permit the detection of single bacterial cells.

Environmental analysis using reporter genes is now a practical technique; multiple reporters have been inserted into a number of pesticide-degrading microbes to permit efficient monitoring after release into the environment.

3. Bioremediation

Air and off-gas treatment

Filterbeds and biofilters

Biological treatment has been used for many years to control odours and volatile organic chemicals (VOCs) in contaminated air. By the 1960s compost and soil beds were being used to control odours (hydrogen sulphide, terpenes, mercaptans, etc.) at sewage treatment plants. The traditional methods of off-gas treatment such as incineration, dispersion, catalytic oxidation, scrubbing and adsorption are best suited to large volumes of well-defined waste gases. Mal-odour problems, from waste plants in particular, are normally caused by complex and varying mixtures at very low concentrations: many of the mercaptans, for example, have odour thresholds below 1 ppb (parts per billion). Biological control of these complex problems offers a simpler alternative to chemical oxidation (Case Study 2). Other advantages of the biological nature of the process are that there are no chemical costs and residues and energy requirements are low.

Case Study 2:   Successful applications of biofiltration – The Netherlands

Since 1989, biofiltration has been successfully applied at MERVO-Hengelo, a soybean and cattle bean toaster. The odours released in the process are eliminated from the off-gases by more than 95 per cent. The filter, packed with Vamfil (a proprietary filtration material), covers an area of 240 m² and is loaded with 300 m³ off-gas per m³ packing per hour. After 2.5 years of operation, the fill material was renewed to prevent too high a pressure drop.

At Promivi (extrusion of cattle feed) in Zwolle, a large part of the odour from the off-gas is eliminated by biofiltration. The filter, filled with Vamfil, has been in operation for two years. Odour elimination efficiency is better than 99 per cent, while the volumetric load is more than 600 m³ off-gas per m³ packing per hour.

In 1989, a biofilter for the abatement of odour emission was sited at a company in Germany specialising in animal rendering. Here, one of the largest gas flows ever applied in biofiltration is still being treated: 214 000 m³/hour. The filter is filled with peat and heather and has a volume of 3 240 m³. An odour reduction of 94-99 per cent was measured by olfactory techniques. The fill material will not have to be renewed until 1992.

Another successful application of biofiltration can be found at a ceramics factory in southern Germany. As a result of drying the products, alcohols are released at a concentration of 230 ppm organic carbon at a flow rate of 30 000 m³ (90 per cent ethanol and 10 per cent isopropanol). A simple biofilter with a volume of 200 m³ was able to remove more than 99 per cent of the alcohols.

Source: Van Groenestijn, Roos and Hesselink, TNO, The Netherlands.

The operation of filterbeds for treatment of contaminated air is analogous to the use of trickling filters for the treatment of waste water. In both applications, bacteria immobilised on support media, under conditions of appropriate pH, temperature, nutrient and oxygen concentrations, degrade organic compounds to their mineral components. For trickling filters, the process cleans contaminated water; for biofilters, it cleans contaminated air.

The basis of biological gas clean-up is the establishment of an active population of micro-organisms on a support medium in which the pollutants are first absorbed and then metabolised. In the traditional system, the support is an extended bed consisting of a mixture of peat and heather. The ideal characteristics of the support have been developed through long experience; the peat heather mix provides nutrients for microbial growth, water retention, a high surface area and sufficient interconnecting air spaces (voidage) to prevent short-cutting by the gas. The heather is added to the peat to give structural support and maintain the voidage over several years of use.

A typical bed may be several hundred square metres in area and will reduce gas components by as much as two orders of magnitude. In practice, performance parameters limit bed depth to 1.0-1.5m. Maintenance costs are low; the bed must be kept wet and drained and needs to be loosened annually. The peat may need to be topped up to keep a constant bed depth. Occasional addition of lime may enhance performance by neutralising the acidity of the peat and adsorbed materials.

In recent years much process engineering work has been carried out, particularly in the Netherlands and Germany. The effort has primarily been devoted to developing biofilters and also bioscrubbers and bio-trickling filters. Biofilters are now applied on a much broader scale, in order to reduce not only odours but also undesirable components such as aromatic and chlorinated organic compounds, ammonia and hydrogen sulphide (Table 2). The technology of biofilters is described in Appendix 5.

The essential features of state of the art commercial biofilters are the nature of the bed, the organisms, and the housing and control systems. Special packing materials with optimal structural properties have led to the reduction of aging phenomena, decrease in flow resistance, and increased lifespan both of the bed and its microbial population. Modern filters, in addition to peat, bark and wood chips, may contain porous clay and polystyrene beads to optimise durability and reactive surface and to reduce back pressure. A proprietary filter with a bed life of five years has permitted pollutant loadings ten times higher than for a traditional filter. Most industrial sources of pollution are not continuous, but biofilters have been shown, to survive for at least two weeks without significant loss of microbial activity.

Treatment of waste gases containing easily biodegradable, xenobiotic compounds may advantageously take place in a multi-staged biofilter. Most new biofilters are closed, thus protecting the filter from adverse weather conditions. They may be in the form of a tower, which, like a carbon adsorption column, may include one or more filter stages with individual water distribution systems for each (Figure 2). The biotower supports substantially higher air loading and has a much smaller footprint than the classical filter bed. Space requirements for towers are thus lower than those for conventional filterbeds and control is more easily effected, but capital costs are higher. Biofilters can compete on a cost basis with physical and chemical process alternatives. However, they are best used for low concentration gas streams, odours and room, rather than process, ventilation.

While biofiltration has, in the past, been regarded as a stand-alone technique, it is likely that combinations of biological and non-biological treatments will prove superior from a cost and efficiency viewpoint. Thus, electrostatic dust precipitators may be used to remove dust and chemical scrubbers to remove gases such as ammonia, ahead of the biofilter which can then remove volatile organics more efficiently. Chemical absorption

Table 2. **Recent biofilter installations**

Pollutant	Industry type
Odour	Animal rendering
Odour	Tobacco
VOCs	Fish processing
Hydrogen sulphide	Landfills
Aromatics	Foundry
Phenol	Phenolic resins
Formaldehyde	Plywood products

Source: P. Hesselink, TNO, Delft, the Netherlands.

Figure 2. **A multi-layer biofilter designed for a Swiss chemical company**

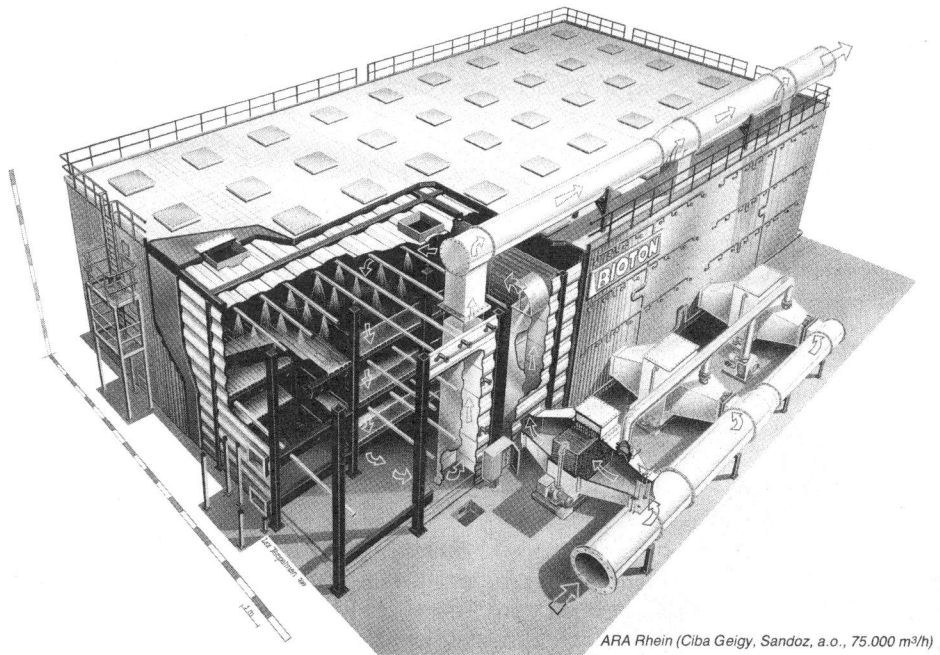

ARA Rhein (Ciba Geigy, Sandoz, a.o., 75.000 m³/h)

Source: Clairtech, Utrecht, the Netherlands.

followed by biological regeneration of the absorbent is the principle of a technique for treating gases containing hydrogen sulphide.

Other techniques

Research and development are currently being undertaken on bioscrubbers and bio-trickling filters. A bioscrubber consists of a packed bed for absorption of the pollutants and a separate tank for biological reactions. Pollutants dissolved in the water are carried to the reaction tank and water is recirculated and nutrients are added to the tank (Case Study 3). A bio-trickling filter is similar but does not have the separate tank because the micro-organisms are immobilised on the packing material and the water stream flows counter-current to the gas stream. Prototype systems have used existing percolating filters for scrubbing sludge gases. The liquid effluent ensures a good biological film. Completely novel chemicals may be broken down after acclimatation and selection. Both techniques may be more suitable for chlorinated hydrocarbons but are still at pilot plant stage and require scale-up to commercial systems.

Case Study 3: Hydrogen sulphide removal using a bioreactor – Japan

The Bio-SR process is a unique process for hydrogen sulphide removal (desulphurisation) developed by Dowa Mining Co., Limited in Japan, which utilises bacteria to regenerate the gas-treating solution following absorption.

Feed gas is brought into contact with acidic ferric sulphate solution in a column, and the sulphide is oxidised to elemental sulphur while reducing the ferric to ferrous sulphate. After separation, the solution is treated in a bioreactor with the bacterium *Thiobacillus ferrooxidans* in the presence of sulphuric acid (pH 2-3) which reoxidises the ferrous sulphate to ferric. The system has no effluent and low energy consumption since the process takes place at atmospheric pressure and ambient temperature. No side reactions take place to degrade the absorbent.

The process has been demonstrated in a sewage treatment plant where concentrations of hydrogen sulphide have been reduced from 400-2 000 ppm to under 10 ppm. A commercial plant has been built for the Barium Chemical Company of Japan which has better than 99.9 per cent recovery of sulphur (150 tonnes per month).

Source: NKK Corporation, Tokyo, Japan.

Soil and land treatment

The term "soil" can be used to describe a variety of physically and chemically diverse media that have formed above widely differing rock materials in different climates, and support various forms of animal and plant life (Figure 3).

The application of organic chemicals to soil can easily perturb soil equilibria, by causing toxic effects on the soil biota and by creating an imbalance in the carbon budget of the soil. Thus, while an agricultural soil is capable of degrading and incorporating inputs of organic farm wastes into microbial biomass, plant nutrients and soil organic matter, it may not break down anthropogenic organics such as oils, tars and chlorinated solvents, without the stimulation of specific micro-organisms and their concomitant enzyme systems.

Contamination of soil can be due to the presence of both organic and inorganic pollutants. Inorganic pollutants range from heavy metals to anions such as sulphate, while organic pollution extends from gross contamination of manufacturing sites (chemical plants, gas works, etc.) on the one hand, to trace pesticide contamination due to agriculture. Pollution can be acute, as when there is a spill, or chronic (*e.g.* leaking

Figure 3. **A typical soil particle. The pores contain air, water, bacteria and organic materials**

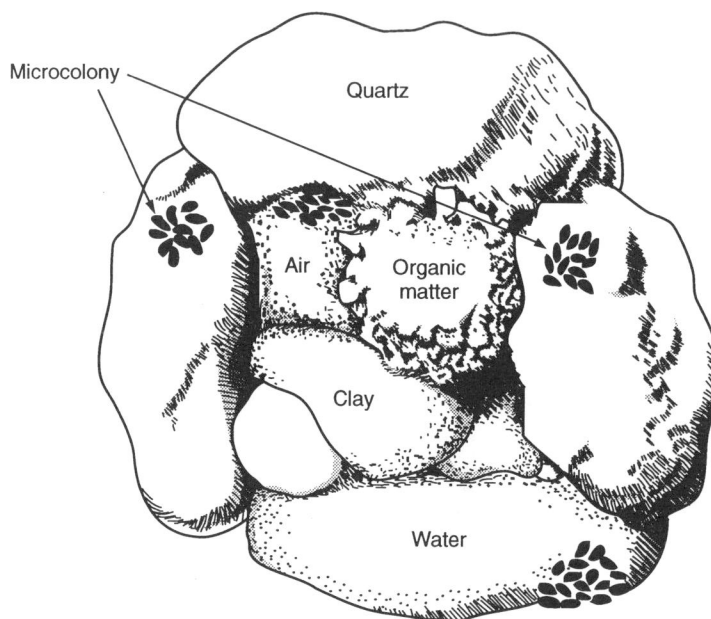

Source: R. Atlas, University of Louisville, United States.

underground storage tanks). The main impact of biotechnology on contaminated soil clean-up to date is under circumstances of organic pollution, which is more open to microbial attack (Table 3). Use of micro-organisms to remove inorganic and especially metallic pollutants is the subject of considerable research. Soil biotreatment is described more fully in Appendix 6.

The principle underlying biodegradation as a tool in soil clean-up is to bring suitable micro-organisms, various essential nutrients (nitrogen, phosphorus, etc.) and, where necessary, air or oxygen, into contact with the polluting material so as to optimise the conditions for breakdown. The micro-organisms will then use the pollutant as a substrate for growth converting it into microbial biomass. It is important to realise that biological treatments, either *in situ* or *ex situ*, result in exponential degradation of the pollutants rather than straight line reduction. Thus, zero levels of pollution are impossible to achieve, but the time to achieve a specific low level of contamination can be predicted.

Table 3. **Chemicals potentially suitable for bioremediation**

Class	Example	Aerobic process	Anarobic process
Monochlorinated aromatic compounds	Chlorobenzene	◆	◆
Benzene, toluene, xylene		◆	◆
Non-halogenated phenolics and cresols	2-methyl-phenol	◆	◆
Polynuclear aromatic hydrocarbons	Creosote	◆	
Alkanes and alkenes	Fuel oil	◆	
Polychlorinated biphenyls	Trichlorobiphenyl	◆	
Chlorophenols	Pentachlorophenol	◆	◆
Nitrogen heterocyclics	Pyridine	◆	
Chlorinated solvents			
– Alkanes	Chloroform	◆	◆
– Alkenes	Trichloroethylene	◆	◆

Source: Environmental Protection Agency, United States.

Problems arise from the mechanical difficulties encountered in trying to manipulate large volumes of soil to get optimum conditions for microbial activity. Also, there is a huge variety of potentially polluting compounds, some of which are relatively easy (*e.g.* petroleum hydrocarbons) and others extremely difficult (*e.g.* PCBs) to degrade.

Technologies for bioremediation fall into two groups: *in situ* treatments and *ex situ* treatments.

In situ treatments

In situ treatments for soil and groundwater involve all those techniques used when the soil is not moved (Figure 4). In such cases, the main problem faced by the operator is to distribute nutrients and air through the contaminated area. While the process uses inorganic nutrients relatively slowly, an actively respiring microbial population can use oxygen at a phenomenal rate, and, consequently, the methods used to introduce oxygen are crucial. Techniques include injection through lances, the use of buried pipes, and the circulation of aerated water.

The advantage of *in situ* treatment is that they do not disturb the site. This can have important implications for the future value of the land. Also, if the site is in a sensitive area or is still operating while remediation is taking place, then *in situ* remediation is the only option. However, harmful metabolites produced *in situ*, with higher solubilities than the original pollutants (polyaromatic hydrocarbons converted into phenols, for example), can themselves pollute groundwater. Non-biological technologies such as soil venting are also being developed for in situ soil clean-up, and these are the main competition to biological technologies. A variant, known as "bioventing", has air or oxygen trickled into soil layers at such a rate that natural organisms are encouraged to metabolise pollutants but not so rapidly that volatile pollutants are stripped into the atmosphere (Case Study 4). *In situ* bioremediation should not be considered complete until contamination levels have reached acceptable levels in the groundwater and the soil, since a second clean-up operation may be needed when recontamination of the groundwater occurs through dissolution of residual contamination in the soil.

Figure 4. **Diagram of an *in situ* bioremediation operation**

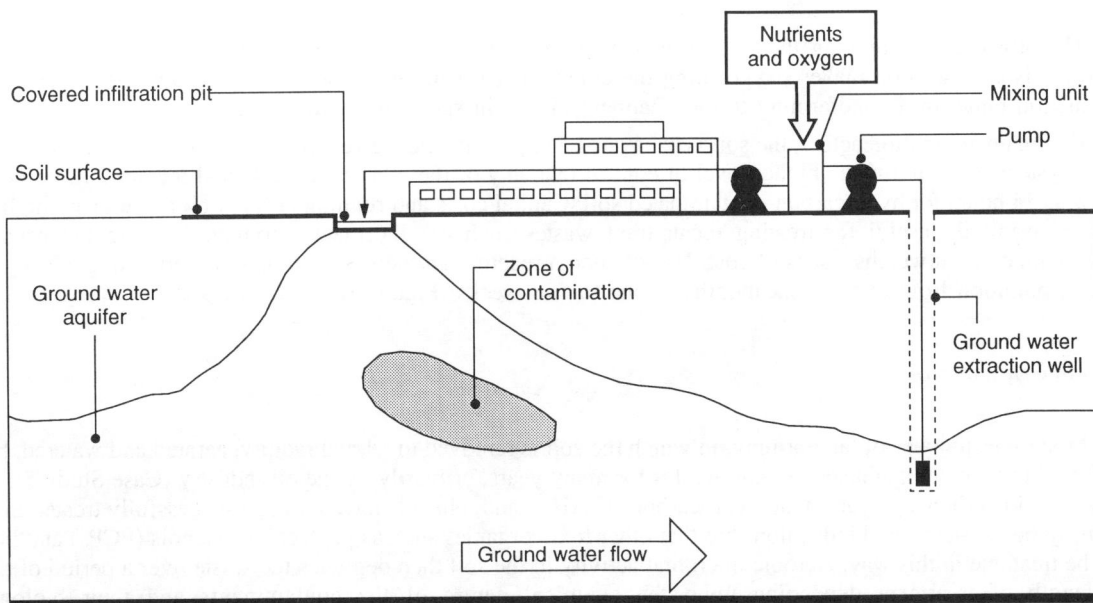

Source: Shell Research Limited, United Kingdom.

Case Study 4:   Soil venting at Hill Air Force Base, Utah – United States

A full-scale soil venting project conducted by the US Air Force to remediate a 27 000 gallon jet fuel spill at Hill Air Force Base, Utah, was recently completed. During this 18-month project, jet fuel residues in soils were reduced from an average total petroleum hydrocarbon (TPH) concentration of approximately 900 mg/kg to less than 10 mg/kg. Monitoring of vented soil gas indicated that volatilisation accounted for 60 per cent of the removal, and biodegradation accounted for the remaining 40 per cent.

In addition to oxygen supply, the distribution of petroleum-degrading bacteria, nutrients and soil moisture conditions influence the success of *in situ* biodegradation. Microbial populations were established at depths of 50 feet prior to initiation of the test. The demonstration indicated that soil bacteria are able to recycle essential nutrients and may also rely on nitrogenase bacteria to fix atmospheric nitrogen and to introduce useful forms of nitrogen for fuel-degrading microbes. Nutrients added to the site produced little or no increase in hydrocarbon degradation rates.

A column test using soils from the Hill AFB site showed increasing fuel biodegradation as soil moisture was increased from 6 to 18 per cent by weight. At higher moisture levels, a reduction in air permeability can limit oxygen supply.

Source: Engineering-Science Inc, United States.

In another technique, known as "air sparging", air injected into the saturated zone beneath the zone of contamination causes volatile organic compounds (VOCs) to partition from the dissolved or sorbed phase into the air phase, where they are transported into the vadose zone as soil vapours and captured and removed by a vapour collection system. Air sparging has been used in the enhanced remediation of gasoline-contaminated saturated soils and groundwater. Air sparging also elevates the dissolved oxygen concentration in the groundwater, enhancing the biodegradation of less volatile, higher molecular-weight compounds.

Ex situ treatments

The advantage of digging the soil up and treating it above ground is that the system is much easier to control. Also, excavation makes oxygenating the contaminated soil very much easier. Technologies for *ex situ* remediation range from landfarming to the treatment of soil in specialised bioreactors.

For treatment in bioreactors, the soil is excavated and put into closed vessels, usually in the form of a slurry. These systems exhibit a much higher level of control than any of the other treatments and retention times can be as low as 24 hours for hydrocarbon-contaminated soil. Capital costs and running costs of bioreactors are high and they are primarily useful for treating recalcitrant wastes such as chlorinated aromatics, where the need for decontamination outweighs issues of cost. Use of such a reactor can reduce the time required to degrade a given level of pollution from three to six months to one to two weeks (Figure 5).

Modern landfarming

Modern techniques of landfarming, in which the soil is removed to a lined facility, aerated and watered, have been used to treat contaminated organic wastes for many years, primarily by the oil industry (Case Study 5). Oil-derived hydrocarbons, polyaromatic hydrocarbons (PAHs), and, phenols have all been successfully treated in soil treatment bed systems, and indications are that other toxic organics such as pentachlorophenols (PCPs) and PCBs may be treatable in this way. Aerobic microbial activity in the soil then degrades the waste over a period of time, the length of treatment depending upon the chemical nature of the contaminants and their biological degradability, as well as on the climatic conditions. Treatment time scales are typically of the order of one or two spring and summer seasons in northern Europe, although significantly faster rates of degradation can be achieved in warmer climates.

Case Study 5: Soil rehabilitation following pipeline damage in Dinslaken Nord – Germany

Owing to a mining subsidence, a leakage occurred in a pipeline supplying the big oil refineries in Germany with petrol. Approximately 275 000 litres of petrol leaked out of the pipeline and seeped into the soil. As a result of the leakage and the surface run-off, 21 000 tons of soil were contaminated.

The firm Umweltschutz Nord was charged with taking care of the soil excavation work and subsequent soil rehabilitation. In order to secure the soil, the area envisaged for storage was sealed underground by welded hydrocarbon-resistant films with a drainage layer. Specially sealed lorries transported 21 300 tons of soil to the site and deposited it there while adhering to the permissible limits for air pollution.

The microbiological soil rehabilitation activities were carried out in accordance with the TERRAFERM BIOSYSTEM ERDE technique. Under this system, the soil is homogeneously mixed with a biosubstrate using special machines of the ''mole'' type and then rehabilitated microbiologically in tents. The biosubstrate acts as a carrier for the micro-organisms that degrade the hydrocarbons. At the same time, nutrients are added to the soil along with the biosubstrate. A special machine continuously extracts air from the contaminated soil by suction and passes it through biofilters. In addition, the composition of the air in the innermost part of the contaminated soil is constantly monitored in order to introduce appropriate remedial measures should the air mixture become explosive.

The goal of rehabilitating the contaminated soil was achieved after a microbial degradation phase of approximately five months.

Source: Umweltschutz Nord, Germany.

Figure 5. **Comparison of biodegradation of pollutants in a bioreactor and by landfarming**

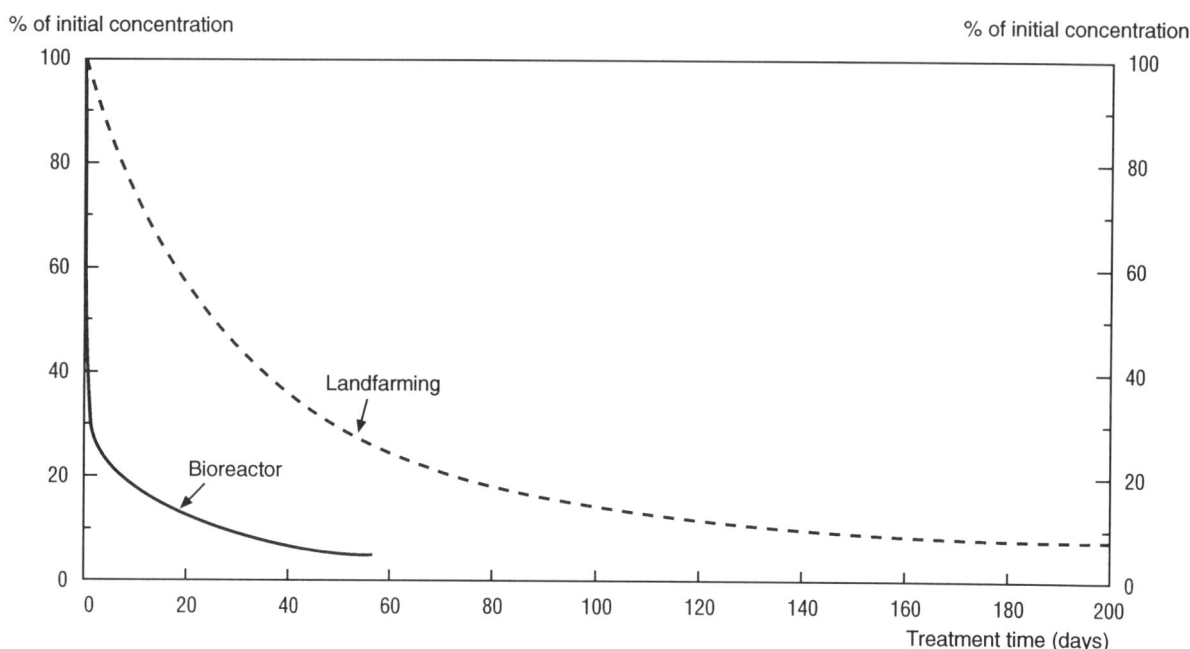

% of initial concentration   % of initial concentration

(Graph: x-axis "Treatment time (days)" from 0 to 200; y-axis "% of initial concentration" from 0 to 100 on both left and right. Two curves labelled "Landfarming" (dashed) and "Bioreactor" (solid).)

Source: TNO, Pipeldoorn, The Netherlands.

It is usually necessary to add nutrient-containing fertilisers (principally nitrogen, phosphorus, and trace elements) to the soil in order to optimise conditions for microbial activity and to facilitate adequate degradation of the contaminants. In addition, the application of specific contaminant-degrading micro-organisms to the soil may enhance rates of degradation, especially in the early stages of a treatment programme, where it is necessary to establish an appropriate population of micro-organisms or where known recalcitrant molecules are present.

Alternative technologies

The technology of bioremediation of contaminated land is beginning to come of age. There have been enough successes both in the United States and Europe to validate a number of the techniques described. In the United States, some of the best examples arise from "Superfund" sites and in Holland, one company, using both biological and non-biological techniques, has the capacity to treat 100 000 tonnes of contaminated soil per year.

It is clear that no one method will provide an answer and that all the techniques have their place. In terms of competition with other technologies, however, bioremediation has the great advantage that in most situations it is the cheapest option, aside from taking the contaminated material to landfill. Legislation in the United States restricting land disposal of hazardous waste and the financial and legal liabilities arising from Superfund have made it possible to develop sophisticated bioremediation technology, and a situation is developing in Germany, Holland and Scandinavia. In the United Kingdom it is still cheaper to take contaminated material to landfill, a situation that is, however, changing rapidly.

The essential requirement for any remediation process is that it should be reliable in terms of the results to be achieved and predictable. Polluted sites are very complex, and the choice of technology is therefore very site-specific. Bioremediation, as the new technology, is in a difficult situation: results from treated sites are required to show that it is predictable; at the same time, contractors hesitate to use it until its reliability is proven. Until this situation changes, the site characterisation costs, although they apply to all remediation technologies, may be particularly disadvantageous to bioremediation methods.

The main problems with bioremediation as a technology are that it is time-consuming, thereby tying up capital and preventing land reuse, and that there is a lack, highlighted in Chapter 3, of scientific understanding of, for example, degradative pathways (Table 4). Although virtually all organic materials are degradable to some degree, the ease of breakdown can vary radically, and consequently much work is needed to isolate strains of

Table 4. **Advantages and disadvantages of bioremediation**

Advantages

An ecologically sound, "natural" process
Destroys rather than transfers contaminants to other media
Usually less expensive than alternatives
Can often be accomplished where the problem is located

Disadvantages

Research is needed to develop appropriate technologies
Often takes longer than other remedial actions
By-products which may be toxic can sometimes be formed

Source: Environmental Protection Agency, United States.

micro-organisms to degrade recalcitrant pollutants to harmless end-products. Biotechnology does have the advantage that because micro-organisms are used to break down the organic matter, the end products are minerals, CO_2, water, and biomass. The most important other remediation technology that achieves a similar breakdown is incineration; all other technologies concentrate the material without changing its form.

Biological treatment of solid wastes

The purpose of solid waste treatment is to achieve a safer, less toxic, more stable material which can be used, re-used, disposed of, or discharged for further treatment. Treatment methods should provide a contained environment which accelerates and controls the rate at which waste materials are broken down. As acceptable discharge levels continue their downward trend, it will become increasingly important to maximise treatment efficiency. This will require a reappraisal of reactor equipment, microbes, and the operating conditions of the waste treatment plant used.

Landfilling

Municipal solid waste (garbage) consists of inorganic and organic material, waste paper, garden refuse, and such industrial refuse as the municipalities will accept. In its simplest form, the treatment is to deposit it as a whole in landfills. In some countries, schemes exist either to encourage householders to separate their waste into inorganic and organic, recyclable (paper and glass), and non-recyclable materials or to have central separation plants. In the latter case, metals and plastics can also be separated. Non-usable material then reaches the landfill, and a major fraction of the organic material may be composted or digested anaerobically.

The waste is dumped in natural or excavated pits, successive layers being covered with intervening layers of earth and some of the wet material decaying under microbial attack. Once the material is compacted in place, particularly when it is covered by an earth layer and fresh air is barred from further access, oxygen buried with the garbage is rapidly used up and the system becomes anaerobic. As the refuse decays, the mass settles and compacts further, allowing additional dumping at the same site.

In poorly managed landfill operations, problems sometimes arise from the formation of biogas and acid leachates, liquids which may cause extensive environmental damage by polluting groundwater if they are allowed to seep from the base of the landfill. In new sites these difficulties are generally overcome by undersealing the landfill with clay or butyl rubber and installing drainage systems coupled with leachate treatment facilities.

Landfills can be an important source of biogas in many industrialised countries. Channels of gravel or perforated pipe are inserted in the landfill to assist controlled migration and to avoid the dangers of explosion (when the gas becomes mixed with air) and of the release of potentially toxic vapour emissions. A typical landfill site might be expected to produce gas for at least six years. The gas typically contains between 40 and 70 per cent methane and has therefore, in its produced form, a calorific value of approximately half that of natural gas. In addition to its economic value, a further reason for collecting the biogas is that methane is a major greenhouse gas. Frequently, landfill-derived biogas is produced for dedicated end uses, particularly the firing of cement and brick kilns. However, as landfill sizes increase and other end uses for the gas are envisaged, extensive calorific value upgrading, together with procedures for the removal of moisture, carbon dioxide, hydrogen sulphide and halogenated hydrocarbon, will become necessary.

Anaerobic digestion of solids

Anaerobic digestion of wastes is increasingly being carried out in preference to aerobic processes such as composting in order to convert the organic content of wastes to usable methane. Although this will inevitably be burned and thereby create carbon dioxide, it diminishes the use of fossil fuel. Conventional temperatures for digestion (25-35 °C) are not high enough to destroy pathogens. However, newer technologies involve thermophilic digestion at 55 °C, and it is accepted that this temperature, held for 24 hours or more, is sufficient to destroy most pathogenic organisms.

The solid residue (humus) from anaerobic digestion is less well characterised than compost (see below), and the process may be more capital-intensive. Nonetheless, development continues on both single and multi-stage processes for solids digestion. An interesting reactor being piloted in the Netherlands has a first stage which renders the organic waste more soluble and a second stage which sets out to mimic the cow's rumen.

Animal manure is also treated by anaerobic digestion. In on-farm plants the purpose may be solely to reduce solids and to produce locally usable biogas. In a modern industrial-scale plant, the sludge is dried after digestion and combined and pelleted with sulphur from an off-gas desulphurisation step and the evaporate from a water clean-up stage. The combined material is then marketed as fertiliser. Conventional anaerobic digestion of manures and sludges takes place in completely stirred vessels at solids concentrations ranging from 3 to 8 per cent. A novel solid state process known as the DRANCO (Dry Anaerobic Composting) process has been developed in Belgium; it operates at a solids concentration of between 30 and 35 per cent and at 55 °C. This reactor can achieve a ten-fold increase in the rate of removal of COD as compared with the traditional process (Case Study 6).

The relative importance of anaerobic digestion for waste control, versus the generation of methane as a fuel, varies in different locations and is dependent on the price and accessibility of alternative fuels, the ability of a population to buy that fuel and the availability of feedstock. In developed countries, disposal of waste may be more important, with gas generation a useful and significant by-product. In poor countries, with proportionately large agricultural sectors and high numbers of farm animals generating manure, extensive use is made of simple anaerobic fermenters in rural areas to produce biogas for cooking, heating and as fuel for small electricity generators.

Composting

Composting is an aerobic process which relies on the structural support of the solids to maintain openness and ventilation during the process. Straw, wood chips, shavings, bark and nut husks are commonly used to provide the support material. Straw, for example, will absorb three times its own weight of water and is commonly co-composted with liquid wastes such as sewage sludge and farm slurry. The benefits of composting are similar to those of anaerobic digestion, that is, the controlled decomposition and stabilisation of organic waste, odour control and, additionally, the destruction or inactivation of pathogens.

Different techniques of composting have been developed for different types of waste. The simplest technique is the open static pile. An alternative method, the windrow process, requires the waste to be arranged in long rows and the stack aerated by periodic mixing using special turning machines (Figure 6). Traditional composting is a batch process, and there is a succession from mesophilic to thermophilic ecology. The heat generated during the rapid growth of the bacteria breaking down the smaller organic molecules raises the temperature to 40-50 °C, when a thermophilic population becomes established. Composting proceeds most rapidly at 55-65 °C. Higher temperatures destroy most of the thermophilic bacteria, leaving only the few that are able to survive at extreme temperatures. Exposure to temperatures above 60 °C for several hours is sufficient to pasteurise the compost, *i.e.* kill off pathogens, according to the US EPA. This temperature can be sustained in an active pile for several days. Higher temperatures are not necessary to remove pathogens and require longer times to re-colonise the pile with the micro-organisms that are active during maturation.

Bioreactor composting systems have been used for domestic refuse in an attempt to reduce the land required for pile and row composting. Bioreactors are easier to control but expensive. There are two basic types, an inclined rotating tube or a standard stirred tank. Residence time in the reactor is five to ten days, but has to be followed by a maturation period of several weeks outside the reactor. New techniques seek to make compost making continuous. In the tunnel system, a moving floor can allow fresh waste material to be laid down at one end and compost removed at the other. In reactor systems, compost can be removed at the walls or floor while waste is added at the centre/top.

Case Study 6:   Anaerobic composting of biowaste in Brecht – Belgium

The DRy ANaerobic COmposting (DRANCO) process is a stabilisation process for the treatment of biodegradable organic substrates. These substrates can be the organic fraction from household refuse, source-separated garbage, restaurant wastes, solid or semi-solid industrial organic wastes, etc. The DRANCO process converts these substrates into energy in the form of biogas and a humus-like final product, called Humotex. The whole process takes place in sealed fermentation towers and consequently there is no odour nuisance.

A full scale DRANCO installation with a digester of 800 m³ treating 10 000 tons of source-separated waste per year has been operational since July 1992 at Brecht in Belgium. The source-separated waste is vegetable, fruit, garden and paper waste. After pretreatment, the fraction is introduced into the fermenter, where it is digested for three weeks at a temperature of 55 °C. After three weeks, the digested residue is pressed in order to produce a dewatered cake of about 55 per cent solids. The organics are finally refined and aerobically composted for a duration of about ten days and then are sold as a high-quality soil additive, free of pathogens.

The biodegradable volatile solids introduced into the digester are converted to biogas. This biogas has a methane content of ca. 55 per cent and is directly converted into electricity using a 290kW gas engine. The electricity is used to operate the installation and partially sold.

A second full-scale DRANCO plant in Salzburg, Austria, is under construction and will be operational in November 1993. The plant has a capacity of 20 000 tons of biowaste per year and will cope with all organic waste gathered from the greater Salzburg area, which has approximately 300 000 inhabitants.

Source: Organic Waste Systems, Belgium.

Figure 6. **A modern composting system**

Source: DHV, Amersfooct, the Netherlands.

An important use of compost is as a soil conditioner. The humus and fibre content of compost make it suitable as a soil builder and expander, particularly in heavy clay or light sandy soils. Compost is not a good fertiliser because the combined nutrients are very low and it is too bulky to transport. It is only beginning to be used in agriculture in Europe, but in Japan and the United States there are special projects to promote its use to avoid dependence on chemical fertilisers. Premium uses, such as in landscaping, horticulture and land reclamation, are the most economic uses of compost at present. Peat substitution may be possible but, in contrast to agricultural uses, commercial peat users do **not** want even the low nutrients it contains.

Waste water and industrial effluents

Biological processes have long been used for water treatment. Traditionally, a major aim was the reduction of organic matter in general; specific compounds were not individually targeted. With the increasing importance of cleaning up industrial effluents, this situation is changing, and processes for removing specific pollutants are being developed.

Effluents are variable in strength and flow, and biological treatment can cope with a wide range of load, including some shocks, better than chemical or physical processes, However, most equipment works more efficiently when the flow is controlled, and it is cheaper to balance surges than to design the biological reactor for the maximum load. Balancing tanks to cope, for example, with the accidental release of product are thus an important part of the effluent treatment process.

Processes and reactors

There are both aerobic and anaerobic biological processes for waste water treatment. Aerobic processes are based on the concept that organic compounds in specific water or waste water are used by micro-organisms as carbon and energy sources, and are oxidised by air (oxygen) to carbon dioxide and water. The microbes multiply and are subsequently separated from the water as a sludge, thereby removing BOD. Aerobic biological treatment

48

systems contain a diverse microbial community that includes numerous species of grazing organisms (bacteria, yeasts, fungi, protozoa, etc.).

Anaerobic processes are treatment systems in which micro-organisms do not use oxygen as a terminal electron acceptor. Rather, many anaerobes carry out reactions in which they use carbon substrates as both the energy source and the terminal electron acceptor. While such biochemical routes produce less metabolic energy per unit of substrate consumed than aerobic systems and anaerobic micro-organisms tend to be slower growing, they may degrade carbonaceous pollutants more efficiently. As in aerobic systems, a whole microbial community is used in anaerobic digestion, but anaerobic systems are almost exclusively based on bacterial communities, with some fungi present.

Anaerobic digestion can be considered to occur in three stages. First, complex biochemicals are broken down in a hydrolysis stage. Second, these molecules are converted to short-chain fatty acids, a process termed acidogenesis. Finally, these acids, together with hydrogen produced by "obligate hydrogen- producing acetogens" are converted into methane. This is termed methanogenesis (Figure 7).

Aerobic treatment has been the established technology for low- and medium-strength wastes and also for toxic and recalcitrant molecules. It also permits the removal of nitrogen and phosphorus (although an anaerobic step is also required). On the other hand, aerobic suspended growth systems require more power for aeration, mixing, and recycling of biomass. Anaerobic processes can often be more effective for highly organic wastes, such as food processing waste waters, municipal sludges, and animal husbandry slurries. In these circumstances, aerobic treatment processes are less suitable because they need an ample oxygen supply, they produce large amounts of excess sludge, and they take up a great deal of space. In comparison, anaerobic waste water treatment plants are more compact and represent an elegant method of reducing organic load: carbon compounds are separated as a combustible gas, methane, to and better than 80 per cent may be recovered as a valuable resource (Table 5).

There are essentially two types of reactor system that can be used for biological treatment of water: fixed film and suspended growth. The former is commonly used in the aerobic trickling or percolating filter, while the

Figure 7.   **Anaerobic digestion of organic material**

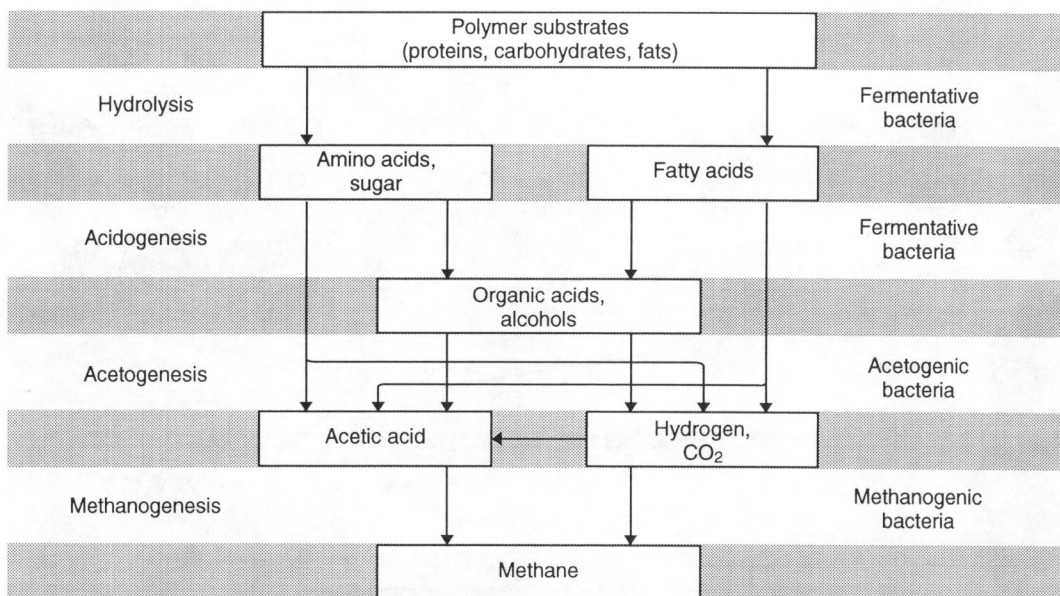

Source: Siemens AG, Germany.

Table 5. **Comparison of aerobic and anaerobic waste water treatment**

Criterion	Treatment	
	Aerobic	Anaerobic
Range of treatable water	Wide	Less wide
Process stability	Good	Less good
Loading	Lower	Higher capacity
Power input	High	Low
Sludge production	High	Lower by factor of 5
BOD and COD removal	To lower levels	
Nitrogen and phosphorus removal	Removes both	Insignificant

Source: W. Verstraete, University of Ghent, Belgium.

latter is used in the anaerobic upflow suspended bed process or the aerobic activated sludge process. A more detailed description of all of these systems is given in Appendix 7.

Fixed film biological waste water treatment systems are often favoured by the water and waste water treatment industry due to their robustness. They can withstand both hydraulic and toxic shock loadings. They are also quite simple to operate and are cheap to run. In addition, reactor systems can employ a relatively short hydraulic residence time, combined with a long solids or biomass retention time. However, fixed film systems require relatively large areas of land per unit of organic loading to be treated and only provide limited efficiency. Trickling filter facilities are also highly seasonal in their operation, with odours in summer and loss of nitrifying

Figure 8. **Comparative size of a biological aerobic flooded filter plant (rectangle) and the equivalent trickling filter (circles)**

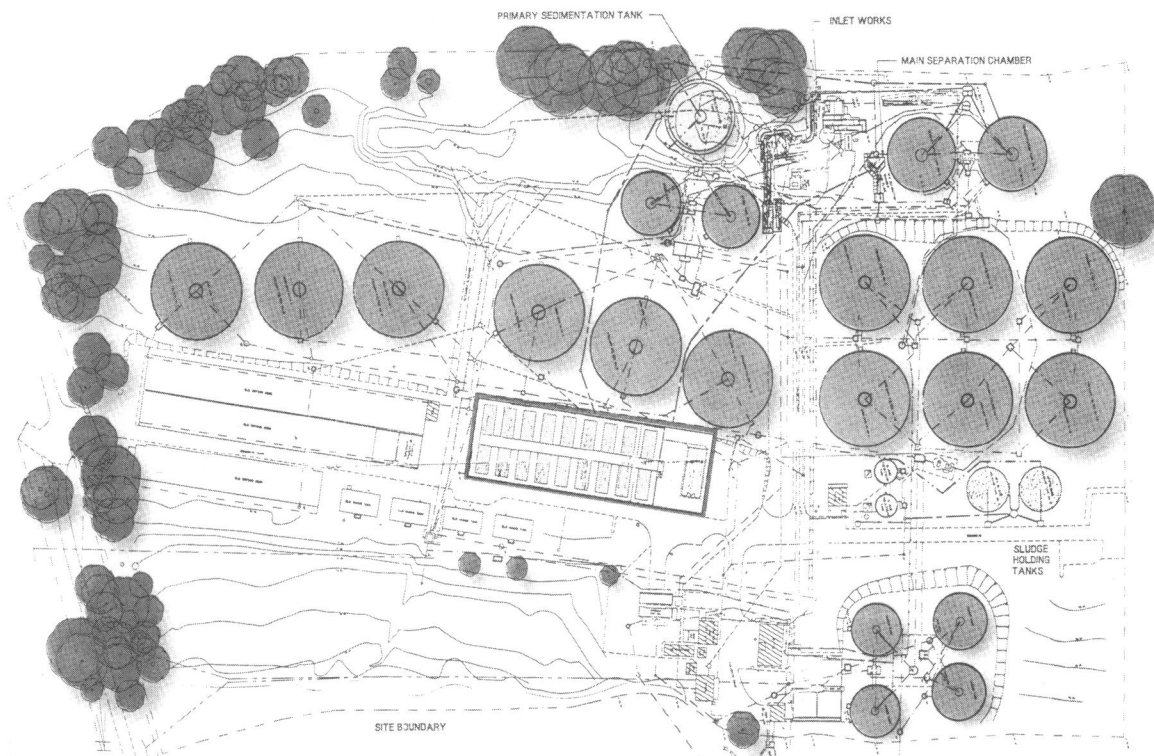

Source: Thames Water, United Kingdom.

(ammonia removal) ability in winter. The system's very robustness is also its Achilles' heel – it is difficult to control.

On the other hand, suspended growth systems can and have been built with enormous capacity: in municipal waste treatment it is not feasible to use fixed film filters for populations of over, say, 50 000. Also, when available space is limited, the ability to build tower, deep-shaft or flooded filter plants is a major advantage (Figure 8).

Suspended growth systems do, however, use more energy in moving liquid about and, in the case of activated sludge processes, in aeration. The latter process is more efficient than percolating filters and is able to treat ten times as much waste. It is therefore cheaper to build but is more difficult to operate and maintain; it also incurs higher running costs due to the mixing and aeration. The largest sewage works in Europe are activated sludge plants. Activated sludge processes are the most widely used for both municipal and industrial waste water clean-up. However, in spite of its importance, the technology still uses variants of a process that was originally developed between 1914 and 1921.

Considerable potential for intensification of both aerobic and anaerobic processes lies in the optimisation of the medium. Most waste water treatment systems are not operated with a medium which is balanced with respect to trace elements, for example. The combination of anaerobic digestion of particulate organics in combination with treatment of soluble organics in a trickling filter offers a further opportunity (Figure 9 and Case Study 7). It reduces sludge formation considerably and also immobilises heavy metals as sulphides.

Nitrogen and phosphorus

Nitrogen removal is increasingly required for waste water biotreatment processes, and appropriate processes are also being introduced into drinking water production. Removal is effected in two stages: "nitrification" of ammonium to nitrate occurs when sludge long enough to prevent washout of the slow-growing nitrifying bacteria; nitrate in turn can be "denitrified" to nitrogen gas by a large group of organisms in the presence of nutrients (BOD) (see Appendix 8). Two-stage sludge processes are sometimes used to remove most of the load as settled sludge in the first stage while maintaining nitrification in the second. Some systems use part of the incoming BOD to denitrify the nitrate formed. In this way, total nitrogen may be removed below limiting levels.

Figure 9. Combination of anaerobic digestion and aerobic treatment for domestic waste water with minimisation of surplus sludge production

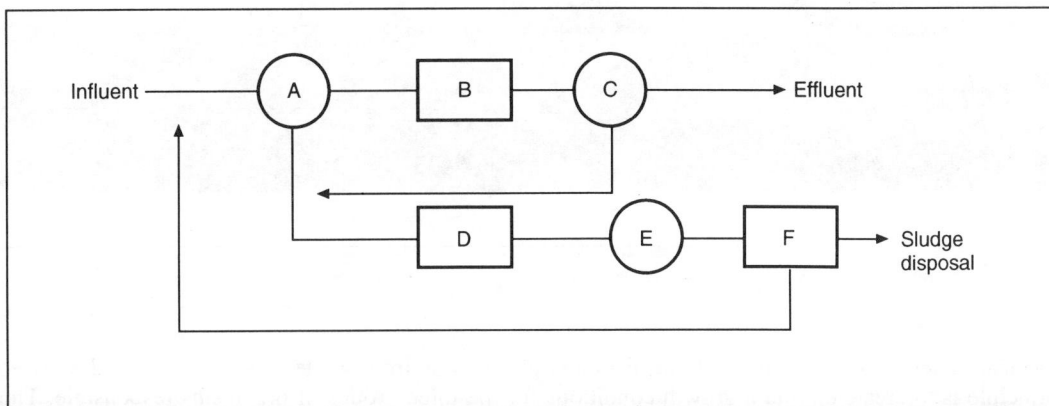

Notes:
A and C: Sludge decantors.
B: Trickling filter or biologically aerated filters.
D: Sludge digester.
E: Sludge thickener
F: Mechanical sludge dewatering.
Source: W. Verstraete, University of Ghent, Belgium.

Case Study 7:   Anaerobic and aerobic degradation of high loaded sulphur-containing waste water, Ringaskiddy – Ireland

The picture shows the anaerobic and aerobic waste water treatment plant at ADM Ringaskiddy, Co. Cork, Ireland, which treats highly concentrated, high sulphate, waste water from a citric and gluconic acid production process. The waste water is first collected and equalised in two large factory effluent holding tanks. It is then conditioned by addition of nutrients and micronutrients, temperature adjustment, and pH control; and it is anaerobically digested in an anaerobic filter-type reactor. The effluent is post-treated in a large activated sludge system with combined nitrification and denitrification. In spite of the sulphur stress, the anaerobic treatment achieves consistently 50 per cent and 80 per cent COD and BOD removal. Energy is recovered through the use of the biogas. The aerobic treatment leads to a removal efficiency of 96 per cent, 68 per cent and 89 per cent for, respectively, BOD, COD and nitrogen.

Source: Biotim, Belgium.

Practical systems for removing both nitrogen and phosphorus from waste waters are being developed. The basic principle is to create optimum growth conditions for the three groups of organisms responsible. Thus, the nitrifying bacteria convert ammonia to nitrate in the *presence* of oxygen, while the denitrifying bacteria convert the nitrate to nitrogen gas, but only in the *absence* of oxygen. Phosphorus-accumulating organisms grow under aerobic conditions but release phosphorus under anaerobic conditions. The challenge is to design plug flow systems in which oxygen levels can be controlled and thus allow the integration of phosphorus and nitrogen removal in the same tank.

An interesting alternative to the removal of phosphorus is the prevention of phosphorus pollution at the source. Thus, use of enzymes in animal feeds can significantly increase the bioavailability of dietary minerals while decreasing the phosphate content of the animal waste by more than 50 per cent.

Concentration

The traditional solution to disposal of sewage effluents has been to rely upon dilution. Receiving waters, such as rivers or the sea, have a certain capacity to tolerate inputs of organic solutions. Multiple uses of water and insistence on high levels of water quality, however, require treatment of even relatively dilute liquid wastes. Environmental legislation is tightening and discharge limits in the ppb range or less are becoming commonplace, especially when large volumes are involved. The extraction of extremely low concentrations of pollutants poses serious problems to engineers. At these levels, conventional treatments such as ion exchange or precipitation are of limited applicability, while activated carbon can be used effectively for organics but less so for heavy metals and is very expensive to regenerate.

One solution to this problem is the use of biological materials to adsorb the contaminants in a manner analogous to activated carbon. The materials used can be plant or animal-derived and can be either living or dead. Reed beds offer an example of a living system adsorbing unwanted material. In this system, the root zone of beds of *Phragmites* reeds provide an ideal environment for microbial populations capable of adsorbing heavy metals and many organic compounds. The system is effective requires a very large area. However, where space is not at a premium this is a low operating cost, environmentally popular, solution. The identification of so-called "hyper-accumulator" plants, which have the property of storing high levels of specific metals is a recent advance. Genetically engineered plants that contain the genes for heavy-metal binding proteins also have been developed, and these may be capable of sequestering toxic pollutants.

Another method, developed in the United States, is to use a bioreactor filled with an inert material such as microporous silica upon which micro-organisms can grow. Water containing trace levels of contaminant is trickled over the inert fill; this causes local concentration increases, thus stimulating microbial activity. Contaminant concentrations have been reduced below detection by this method. It differs from a traditional trickling filter by the level of control of aeration and nutrient addition which it permits.

Where space is tight, it is possible to use the adsorptive capacity of non-living biomass to remove heavy metals and trace organics from effluent and potable waters. Biomass materials effective for this system are usually waste products of the food and pharmaceutical industries and consequently represent a low-cost source material. These materials can remove low levels of contaminant materials selectively, concentrating the contaminant typically by 3-500 fold, and they can be regenerated easily. In the case of heavy metal removal, they can remove toxic metals even in the presence of high salt concentrations.

Biologically based adsorption processes typically are most useful on low level contamination and are best as a polishing step to enhance conventional treatments. Currently, apart from the root zone reed bed system, most adsorbent products are still in the trial stage, although one derived from algae is being sold in the United States under the name "Algasorb". Trials include packed bed systems for removing both the colour caused by peat breakdown products from drinking water and heavy metals from power station and coal gasification effluents.

Target levels of water contamination are being reduced all the time, particularly in countries like the Netherlands where the water table is very vulnerable. These levels are unlikely to be achievable in a conventional treatment system because concentrations will drop below those needed to sustain microbial metabolism. Consequently, a two-stage system is used, with conventional biological treatment followed by an adsorption phase.

All of these methods leave unanswered the question of what to do with the material, organic or inorganic, into which the pollutant has been concentrated. Reduction in volume may make it more suitable for landfill or, in the case of metals, it may be possible to recycle it, treating the concentrate as a raw material.

Degradation

The increasing number of xenobiotic organic compounds daily entering the environment has meant that biotechnology is being specifically targeted at the degradation of complex organic molecules. The advantage of degradative systems is that the contaminant is broken down into harmless end products; the disadvantage is that they are only truly effective on organic materials (Case Study 8). It should be borne in mind, however, that where degradation is incomplete, the process may give rise to residues that are more recalcitrant than the original pollutants. This may be the case for some traditional pesticides, for example.

Source: Environmental Protection Agency, United States.

Novel technologies are being developed for treating recalcitrant compounds. One approach is to analyse the different components of an effluent stream and isolate micro-organisms that can degrade them in pure culture. Once organisms have been isolated that can degrade all the components, the strains can be grown as a mixed culture on the effluent. Notable successes have been achieved in the treatment of wastes from acrylonitrile manufacture using this technique where the only, very expensive, alternative was to bury the effluent in deep faults under high pressure.

Enzymes have been applied to detoxify waste streams, as in the use of parathion hydrolase to degrade the pesticide, parathion. However, detoxification is not equivalent to purification and such systems are therefore only pretreatments. Food treatment plants and restaurants use lipases, particularly, to break down fats and greases in order to make them more readily accessible to further treatment in municipal sewage systems. In general, however, since enzymes are expensive, they are unlikely to be widely applied as end of pipe reagents but rather will be used in process-integrated options. Raw extracts of fungi, which have high levels of enzymes, may eventually form the basis for a cost-effective alternative.

Effective systems have been developed that combine physico-chemical treatments with biological treatments; an example is the use of ozonolysis to break up recalcitrant molecules into simpler compounds that can be broken down more easily. This technique is being used in the water industry for breaking down pesticide residues and other trace organics. Following ozone treatment, the water is passed over a trickling filter of active carbon with a biofilm adsorbed to it which breaks down all the fragments (Case Study 9).

Oil spills

Short-chain hydrocarbon molecules in fuels are more susceptible to microbial degradation than are the longer chain compounds characteristic of viscous oils, tars and waxes, but even crude oil may degrade slowly in nature. The rate of degradation of an oil spill is determined by the heavy (long chain) components because, before a natural microbial attack on spilled oil can have much effect, the lighter components may have evaporated. Molecular oxygen appears essential for the biodegradation of oil residues, if it is to proceed at all rapidly.

Oil slicks at sea, after preliminary recovery by mechanical skimming, have often been dispersed into small globules with industrial surfactants in order to promote more rapid degradation by natural microbial populations. However, surfactants tend to encourage the oil to sink before it has been degraded and many are toxic to marine life. They also tend to spread the oil over a wide area so that fish are unable to swim clear and consequently die from lack of oxygen; their use is now discouraged as a result. The slick may be seeded with oil-degrading bacteria accumulated in advance and held in store for just such an eventuality, and on some occasions mineral salts are added if it is felt there is a local deficiency. While there is a view that detergents do more damage than the oil they are meant to disperse, there is a growing interest in the use of biosurfactants as dispersants; because of their own biodegradability, they do not persist in the environment.

Various methods are used to try to clean oil off beaches and rocks once it has come ashore. These include the use of high-pressure water hoses, with or without detergents where permitted, to wash the oil down a beach where it can be collected and removed, as well as spraying with nutrient salts to encourage microbial activity. Oleophilic fertilisers may provide the necessary long-lasting nutrients for oil-degrading bacteria (Case Study 10).

Groundwater and aquifer contamination

As legislation regarding the contamination of groundwater tightens, new methods for treating it are being developed. The traditionally preferred approach has been to pump the groundwater to the surface and treat it there. Volumes tend to be large and because each treatment is likely to be one-off, building a large-scale plant is not feasible. Treatment is by physical separation where possible, with biological treatment of the residue in either a simple lagoon system or a portable bioreactor.

The most ubiquitous contaminants of aquifers and groundwater are non-chlorinated organic solvents such as ethanol, acetone and methylethylketone; solvents and fuel components such as benzene, toluene, ethylbenzene

---

Case Study 9: Physico-chemical and anaerobic treatment of epichlorohydrin-containing waste water, Pisticci – Italy

In the last decades, the petro-chemical industry has consumed increasing quantities of epichlorohydrin *e.g.* for epoxy resins and glycerol production. These products are synthesised under severe conditions (use of alcoholic solvents, alkaline pHs, high temperatures, etc.) and highly polluted, alkaline, saline and toxic waste waters are produced. Waste water toxicity is mainly due to unconverted epichlorohydrin and to intermediates such as glycidol and chlorohydrins.

In the Epoxital plant of Pisticci (Matera, Italy), epoxy resins are produced from epichlorohydrin and phenolic compounds. The synthesis takes place in batches, in the presence of isopryl alcohol as solvent and NaOH as catalyst. Synthesis waste water approaches 60 per cent of the global polluting load generated by the industrial unit; the remaining load comes from resin purification and plant washings. The synthesis effluents, originated at temperature of 80 °C, have the following main characteristics (g/l): COD 40-70, NaCl 260-290, glycerol 6-8, isopropyl alcohol 8-13, glycidol 5-10, epichlorohydrin 4-6.

Eniricerche S.p.A. has recently developed and patented a process for the treatment of epichlorohydrin-containing effluents. The process is based on the following main steps: *i)* alkaline hydrolysis of the toxic/recalcitrant organo-chlorinated molecules; *ii)* sodium chloride removal; and *iii)* anaerobic digestion of the detoxified waste water. The alkaline hydrolysis is performed in order to lower the concentrations of the most toxic molecules down to a level (epichlorohydrin 15 mg/l, glycidol 1500 mg/l) tolerable by a properly selected and acclimatised anaerobic microflora. The hydrolytic treatment exploits the alkalinity and the high thermic level of the waste water. The sodium chloride is removed by crystallisation (recovery yield 80 per cent). The recovered salt, due to its purity (NaCl 95-98 per cent), can be placed in the market of industrial brines. Mother liquors and condensate from crystallization are mixed with waste water from other plants, add proper nutrients and fed to an anaerobic filter. In this bioreactor the microflora converts the organic compounds into biogas.

A demonstration plant based on this process is under construction in Pisticci. Its main characteristics are as follows:

- anaerobic filter operative volume: 150 m^3;
- feeding flow-rate: 2-3 m^3/h;
- COD in: 14-16 kg/m^3;
- COD out: 2-3 kg/m^3;
- biogas: CH$_4$ 60-70% 15-20 m^3/h;
- sodium chloride: 3.6 t/d.

Start-up is scheduled for the beginning of 1994.

Simplified diagram and photograph of the plant for the treatment of epichlorohydrin-containing wastewater

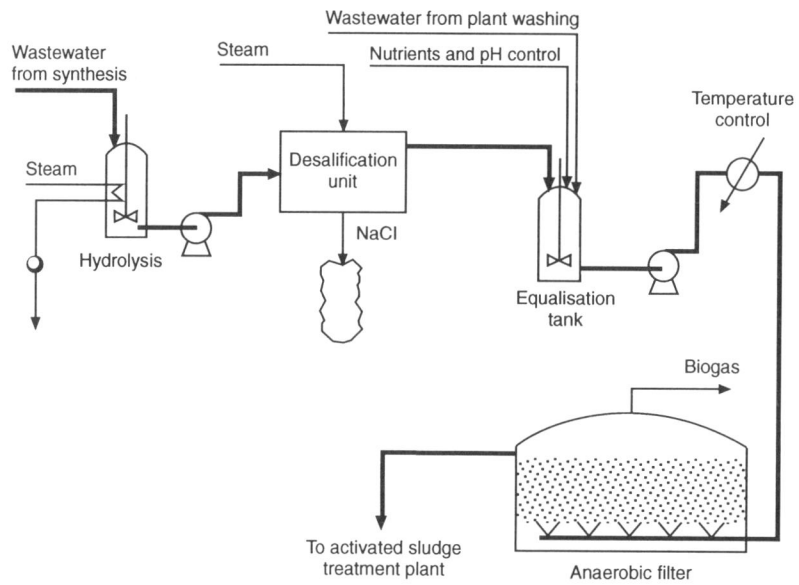

Source: ENIRICERCHE S.p.A.

Source: R. Atlas, University of Louisville, United States.

and xylenes (the BTEX complex); and ethylene glycol. These materials degrade readily in groundwater if adequate oxygen is available. When these materials persist in aquifers, the plume is invariably found to be deoxygenated. Pesticides and halogenated compounds are also increasingly found.

It is difficult to mix oxygenated groundwater into a plume by injection – the injected water simply pushes the plume out of the way – although this has been successful in certain circumstances (Case Study 11). There are two practical solutions: conventionally, water is pumped to the surface, aerated, then reinjected; alternatively, air may be injected into the aquifer at depths below the water table in a process called air sparging (Case Study 12). In the latter, air is injected at a rate that temporarily dewaters an area around the well screen. The dewatered area is buoyant, and rises toward the water table. When the area contacts the water table, the air rushes out through the contact, and the dewatered area collapses. Freshly injected air starts the whole process over again. The buoyant rise drives a convection cell, recruiting deoxygenated water from deep in the aquifer and releasing oxygenated water at the water table. For a single injection well, the total flow field resembles a torus or doughnut shape.

Modelling and field work are needed to establish protocols to estimate proper screen depths, radius of influence of oxygenated water, and appropriate air injection rates. A sophisticated understanding of subsurface processes is important, because it is also possible to cause upwelling of clean water into a monitoring well, giving the false impression that the aquifer has been cleansed, and to send contaminated groundwater on a new trajectory, thereby affecting previously unaffected wells.

Aerobic biosystems show promise for removing the BTEX complex. A combination of nitrate and oxygen promotes biological activity that effectively removes the BTEX compounds from spills of petroleum-derived hydrocarbons; this is likely to be the technology of choice for remediation of fuel spills below the water table.

At present there are two primary options for bioremediation of chlorinated solvents in aquifers. Most of these compounds do not support aerobic microbial metabolism, which explains their persistence. However, they can be co-metabolised by micro-organisms that grow on gaseous alkanes such as methane, butane or propane, by micro-organisms that grow on aromatic organics such as toluene or phenol, and by ammonia oxidising micro-organisms. The physiology and enzymology of the aerobic co-metabolism of chlorinated solvents is now well understood. The efficacy of methane-oxidising micro-organisms has been evaluated at pilot scale but more work is needed in engineering development to make the process a practical alternative to current pump and treat technology. The efficacy of micro-organisms that oxidise aromatic organics has also been evaluated at pilot scale, and results compare well to the methane-oxidising process. The performance of ammonia-oxidising micro-organisms has only been established at bench scale, however.

The second option is microbial reductive dechlorination, which often accompanies anaerobic microbial biodegradation of organics in aquifers contaminated by landfill leachate, sewage, or fuel spills. When it occurs naturally, it is usually considered to be undesirable. The dechlorination products of tetra- and trichloroethene,

Case Study 11:   Bioventing and air sparging in Denver, Colorado – United States

Leaking underground storage tanks at a bus maintenance and fueling facility in Denver, Colorado, had contaminated soil and groundwater with petroleum hydrocarbons. Gasoline, diesel fuel, and/or lubricating oil had contaminated the entire 10 ft thick unsaturated soil profile near the tanks. The petroleum had also percolated to the groundwater table and was transported down-gradient in the groundwater through sand and gravel aquifer material.

The remedial action plan consisted of a bioventing system to reduce petroleum hydrocarbon residues in the unsaturated zone and an air sparging system to remove dissolved hydrocarbons from the groundwater in source areas and prevent further migration of contaminants into the plume.

Bioventing and air sparging pilot tests were completed to better define the volatilisation/biogradation potential at the site; to determine the venting/air sparging well spacing and the vapour collection and air injection rates required to uniformly distribute air (oxygen) through the unsaturated and saturated zones; and to measure the rates of fuel biodegradation by the indigenous bacterial population under oxygen-enhanced conditions. A full-scale bioventing and air sparging system was designed and installed and is currently undergoing startup.

To determine if natural petroleum hydrocarbon biodegradation was occurring in contaminated groundwater, *in situ* measurements of dissolved oxygen (DO) were obtained in monitoring wells installed in the plume areas and in areas unaffected by the hydrocarbon releases. Depressed DO concentrations in petroleum-contaminated groundwater are an excellent indicator of biological petroleum hydrcarbon degradation. DO concentrations in petroleum-contaminated groundwater ranged from 0.0 to 2.8 mg/l, with an average concentration of 0.7 mg/l. These low concentrations strongly indicate that the native bacterial population is capable of degrading benzene, toluene, ethylbenzene, and xylenes (BTEX) and other petroleum hydrocarbons in the presence of available DO. The DO concentrations in uncontaminated groundwater were much higher, ranging from 3.8 to 8.5 mg/l, with an average concentration of 5.4 mg/l. High DO concentrations in uncontaminated groundwater indicate that there are no significant amounts of natural organic material available for bacterial consumption. Thus, DO depletion in the petroleum hydrocarbon plume area can be attributed specifically to petroleum degradation rather than to the consumption of naturally occurring organic compounds in the groundwater. These results established both active bioremediation technologies and passive natural attenuation as potential remedial alternatives for this site.

The remedial actions at this facility demonstrate the advantages of integrating bioventing, air sparging, and natural attenuation modelling for complete site remediation. The *in situ* respiration test indicated that existing bacteria are capable of consuming fuel residues in the vadose zone at a rate of approximately 4 200 mg/kg per year, and the air sparging pilot test illustrated the effectiveness of a single well point in influencing a relatively large volume of groundwater.

The cost of remediation using this integrated approach is expected to be very competitive when compared to traditional approaches. Typical capital and operating costs for site remediation are $7 to $10 per cubic yard of contaminated media for the first year, and an additional $2 per cubic yard for each additional year of operation. This integrated approach offers a simple and cost-competitive solution to many petroleum contamination problems.

Source: Engineering-Science Inc., United States.

particularly vinyl chloride, are often more hazardous than the original pollutant. When dechlorination is complete, however, the ultimate products are environmentally benign, and the process can be used for remediation.

Groundwater may also be cleaned up in above-ground bioreactors (AGBs) (Figure 10). The technology employed is akin to that employed in suspended system waste treatment plants. The waste is presented to the reactor in liquid form. The biomass is normally present in a suspended form, using various support materials including diatomaceous earth, activated carbon, glass, etc. Immobilising the biomass on the supporting substrate confers a dual advantage to the process in that it reduces the amount of sludge produced while increasing the biomass retained and available to transform the waste material(s). The AGB may be operated in an aerobic or anaerobic mode; the aerobic treatment processes has been used most frequently to date. Examples of the successful application of AGB technology include the treatment of pentachlorophenol in contaminated surface and groundwater and organochlorine and organophosphate pesticides from contaminated groundwater.

Case Study 12: Contaminated groundwater in Karlsruhe – Germany

Mineral oil leaks from the marshalling yard of the German Railway System contaminated the groundwater near a municipal water plant. Although the marshalling yard is located in the down-gradient direction of regional groundwater flow from the pumping wells, the cone of depression of these wells extended well beyond the marshalling yard. Organic contaminants (contributed primarily by the well nearest to the marshalling yard) were detected in the well water at the plant within 10 years of installation of the wells. The nearest well was subsequently taken out of service in 1978. Within a few months of the shutdown of the first well, the water quality of the second showed a similar decline. Without the initiation of remedial measures, the contamination would have resulted in the complete shutdown of the water plant.

The contaminant was mineral oil, which contained various hydrocarbon fractions. During the later stages of investigation, complexes of hexacyanoferrate (a food additive) were found in well water. The cyanide complex was leaking from the disposal areas of an old chemical plant which has been shut down decades earlier.

Based upon past experiences of groundwater treatment with ozone and biological purification by activated carbon in other areas of Germany, an ozone treatment system was used for water from the first well, and the ozone-treated water was infiltrated via five injection wells nearby. A further three infiltration wells were installed at a distance of 75m from this well, between it and the marshalling yard.

Ozone was produced on-site. The ozonization system treated 450cu m of water, and the oxygen content of the ozone-treated water was approximately 9 mg/l. The rates of injection through the five infiltration wells varied during the course of remediation. Water levels rose on the order of 0.5 to 0.6m in the vicinity of the injection wells. Through injection and extraction, a hydraulic control was established that retarded further plume migration.

Substantial amounts of oxygen were taken up in the first phase of operation. Thereafter, oxygen consumption from the infiltration was reduced, and the dissolved oxygen content in the groundwater increased. Total microbial cell counts increased by factors of 5 to 10 after the initiation of operation. Dissolved hydrocarbons were completely mineralised from the groundwater after 1.5 years of operation.

Source: Environmental Protection Agency, United States.

Figure 10. **Diagram of a groundwater treatment operation**

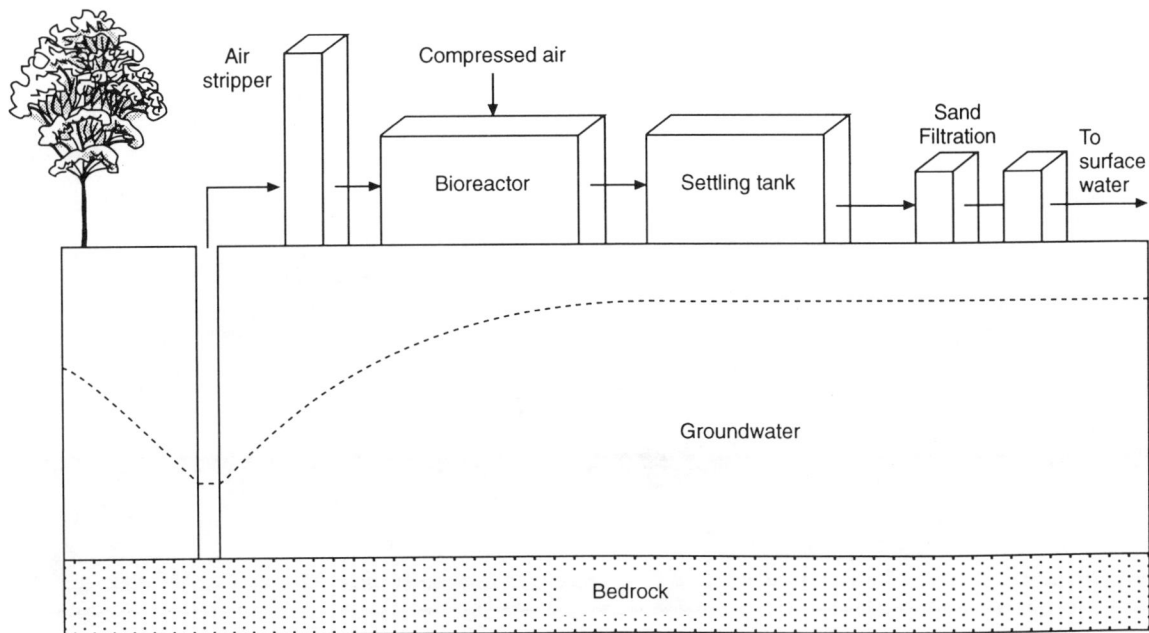

Drinking water

In most developed countries, water may be recycled a number of times before finally reaching the sea or air. Consequently, technology for producing drinking water from waste water is an essential component of environmental clean-up. Biotechnology has an essential role to play; indeed water percolating through the soil ecosystem is bio-purified as chemical and biological contaminants are removed.

In the past, the technique of slow sand filtration of surface and groundwater was an important biological method for providing industrial societies with quality drinking water, and it still is a crucial technique in developing countries. Industrial countries have to rely more and more on contaminated surface water, and, therefore, classical physico-chemical processes, such as coagulation and adsorption, have to be complemented by advanced biological processes that remove contaminants such as nitrate, manganese and assimilable organics in order to achieve the required high quality. In years to come, microbial processes for removing trace pollutants at low and sub-physiological concentrations will become paramount.

In those countries that rely heavily on groundwater as their source of drinking water, techniques for *in situ* remediation, which are further refinements on those used for pollution removal, are being developed. Thus, technology from Sweden involves encouraging naturally occurring bacteria around a main pumping well to precipitate out manganese and iron, effectively filtering the groundwater as it passes through the zone.

Chapter 3

Research and Development – Recommendations

Environmental treatment systems must be in harmony with the global material cycle in the biosphere. They should be gentle to humans and to the environment and have four key characteristics: the renewability of resources; the mildness of production processes; the compatibility of products and waste streams with humans and the environment; and the recyclability of wastes.

In the face of the increasing severity of the environmental damage of this century, there has been a growing awareness on the part of the public and governments that actions are needed to maintain and restore environmental quality. While waste minimisation and recycling programmes have been instituted in many industries and by public authorities to help conserve resources and to protect the environment against the release of wastes and pollutants, new technologies are needed to aid in sustaining development and environmental quality. The accelerated development of biotechnology during the last decade presents new possibilities for dealing with both current and emerging problems.

In the past, biotechnology has provided solutions for environmental needs. During the nineteenth century, when the disposal of wastes into rivers threatened human health and the well-being of aquatic life, waste water treatment facilities were developed as one of the first applications of biotechnology for the maintenance and restoration of environmental quality, and this treatment has been of great benefit to humankind. Since that time, only minor changes have been made in the fundamental design of the original sewage treatment plants and the way organisms are used, and these facilities now often fail to meet performance criteria.

Individual applications of biotechnology have been used for bioremediation of polluted sites *in situ* but biotechnology has not yet become an integral part of overall pollution remediation efforts. Site specificity and the inability to predict and to monitor performance have limited the acceptance of biotechnological solutions by engineers and managers charged with deciding on appropriate environmental remediation solutions.

Biotechnology is one of several available technologies for the maintenance of environmental quality and must be viewed within the larger spectrum of scientific and engineering disciplines. Research needs to address the legacy of historical and present pollution and also the means of preventing future pollution. It should be borne in mind that replacement of chemical with biochemical processes will not necessarily reduce waste; indeed it may create more, but the waste will usually be more manageable. The importance of biotechnology in this overall context has increased significantly within the past five years and will continue to increase. This is due in part to the exceptionally rapid advances of knowledge, the low environmental impact and cost-effectiveness of using biological as opposed to physio-chemical treatments, and also to problems associated with non-biological treatments such as, for example, the generation of gaseous pollutants during the destruction of other pollutants by incineration.

Emerging threats to the environment, such as the atmospheric changes that potentially pose a threat of global climate change, the loss of trees and forests due to atmospheric pollutants, and the formation of deserts, are appropriate topics for biotechnology. Biopolymers, synthetic fuels and other biological alternatives to chemical processes present opportunities for economic and environmental benefit.

Biotechnology is not a panacea for environmental clean-up, nor should the successful development of processes be seen as a licence to continue to pollute.

Some of the research is short-term and some is long-term. However, short or long it all needs to be started as soon as possible.

There are a number of specific R&D bottlenecks and informations gaps which have to be addressed. These are associated with fundamental research and bioprocess engineering and are particularly related to:

Fundamental research:

– molecular biology;

- microbial physiology and metabolism;
- microbial ecology and genetics;

Bioprocess engineering:

- scale-up problems;
- modelling and control theory;
- analysis and measurement;
- reaction kinetics and growth-limiting factors;
- process design and optimisation;
- bioavailability.

1. Molecular biology

The better we can describe and understand complex biological systems and identify the blueprint and the basic units of living organisms, the greater will be our ability to use such systems to develop new technologies. This knowledge will lead to making better use of biological production processes (cells and cell-free systems) for manufacturing, without increased energy consumption, exploitation of raw materials, or increased use of landfilling. Any by-products of manufacturing will, to a large extent, be recycled or will be avoided completely; risks at the workplace will be reduced, and production systems compatible with the environment and which conserve resources will be more favourable to the Earth's ecosystem.

A fundamental understanding is therefore required of the molecular structure of cells and their components to provide a basis for technological development. In the final analysis, it is genome research that will enable these new processes and substances to be developed. Examples of such systems include new biocatalysts, biopolymers and biomembranes developed on the basis of a better understanding of cell membranes.

2. Microbial physiology and metabolism

Although the need for a better understanding of reaction kinetics has been stressed, this is not to imply that knowledge of microbial metabolic pathways is complete. One example lies in the path of bacterial nitrogen metabolism: denitrification processes are generally thought to produce gaseous nitrogen, an environmentally acceptable end product that can be released to the environment without adverse effects. However, there is increasing evidence that intermediates in the denitrification pathway, particularly nitrous oxide, are frequently formed and liberated during the process, and there are alternative pathways leading not to nitrogen but back to ammonia. Nitrous oxide is, of course, one of the gases that is implicated in both the tropospheric ''greenhouse'' effect and in the destruction of the stratospheric ozone layer, so that microbiological processes leading to accidental nitrous oxide production and release into the atmosphere need to be much better understood.

An area which requires a special focus is the metabolism and co-metabolism of recalcitrant organic compounds, particularly polychlorinated hydrocarbons. It is known that organisms able to metabolise certain compounds such as methane and toluene can simultaneously degrade or partially transform other more intractable materials carrying electro-negative groups (halo-, nitro- and sulpho-). Although the metabolic ability to mineralise these compounds frequently exists, the reactions are often poorly induced by their substrates, subject to cross-inhibition by other intermediates, and have metabolites misrouted into dead-end toxic products when mixed substrates are present. The precise nature of the reactions needs further elucidation, and research is needed to improve our understanding of degradative pathways, particularly under process conditions.

3. Microbial ecology and genetics

There exists a lack of knowledge regarding the functioning of complex microbial communities. Little is known about the environmental conditions required for the functioning of an ecosystem and about interactions between micro-organisms (Figure 11). Exploitation of the biological potential requires this knowledge.

There are two separate parts to the study of the ecology of micro-organisms. It is important to know more about the interactions, competitive and collaborative, between organisms in the microcosm of a mixed culture reactor. Equally important, and considerably more difficult to obtain, is information about interactions in soil

Figure 11. **Different levels of ecological organisation**

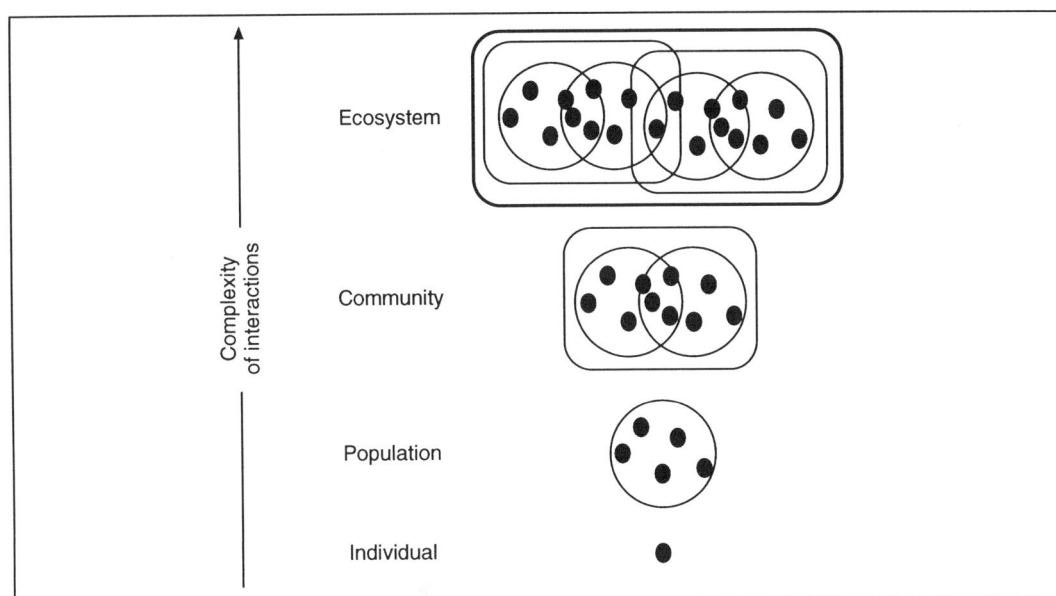

Source: R. Atlas, University of Louisville, United States.

ecosystems. In the latter case, the number of species actively interacting is more diverse and includes earthworms, plants, and man.

It is into this latter environment that GMOs may be released, and it is imperative to be able to follow them (and their genetic material) and to observe their involvement with the natural flora and fauna. Development of new vectors, genetic strategies, and tools are required to ensure the appropriate ecological behaviour of recombinant micro-organisms within predictable limits. Novel non-disruptive techniques are needed to monitor the status of specific strains within complex microbial communities in relation to their biodegradative performance and other factors. Ecological issues involved in field testing of micro-organisms in waste treatment include: the functional role or niche of the micro-organism in the microbial community and ecosystem; the potential for gene exchange between microbial taxa; the potential ecological consequences of the persistence and spread of such micro-organisms and their genetic material; and measures to control, if necessary, the effects of introduced micro-organisms. High concentrations of pollutants are toxic to some microbes and can alter species composition and community diversity.

It is widely accepted that only a small part of the microbial diversity has been mapped. This situation is a consequence both of the lack of support for microbial taxonomy and also the difficulty of culturing many organisms. Remedies include training, identification of new isolation and culture techniques, and a wider screening for degradative function.

4. Scale-up problems

A variety of problems emerge when transferring simple laboratory experiments into the field. In small-scale bioreactors, it is possible to achieve a homogeneous suspension of micro-organisms and uniform concentrations of dissolved nutrients. However, as the process is scaled up, inhomogeneities and concentration gradients can occur, due to increases in circulation time and mixing time caused by the larger volume. These changes may affect the performance of the culture, reducing growth rate and/or productivity. The difficulties encountered with mixing may, in some instances, severely restrict the scale of bioprocessing and ultimately the commercial applicability of bioremediation as a waste treatment option for a given process.

Figure 12. **A typical microcosm for determining the fate of pollutants
in a salt marsh**

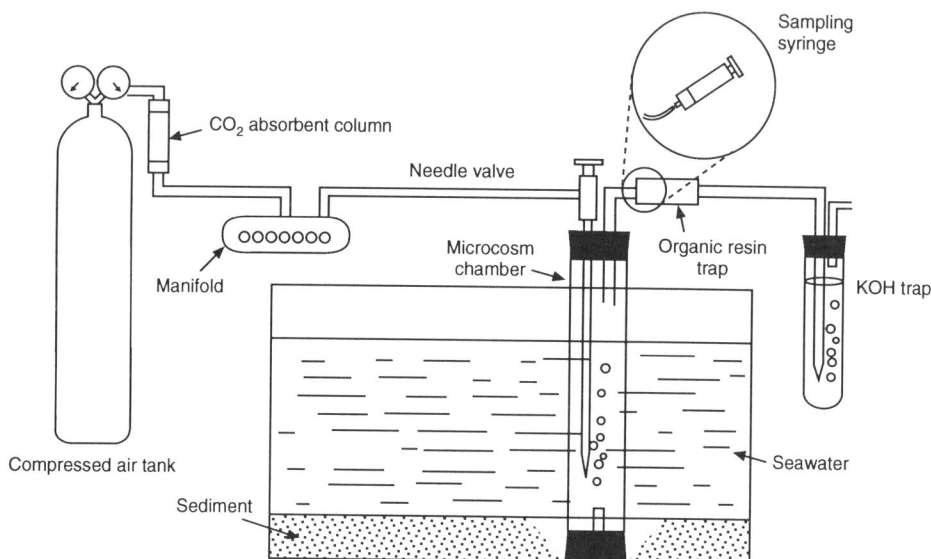

Source: R. Atlas, University of Louisville, United States.

Extrapolations from small operations in the laboratory to field programmes are often complicated by site heterogeneity, and more research at the intermediate or pilot-scale level is required. The availability of small-scale simulations (Figure 12) would facilitate the development of new technologies and permit the better identification of which bioremediation technologies will be useful under field conditions. The design of microcosms which are, at one and the same time, manageable and a realistic simulation of field conditions is essential.

5. Modelling and control theory

A major problem for the effective operation of bioremediation processes is the development of a mathematical approach, with relatively broad applicability, for describing such processes. The development of models that can describe the important physical, biological and molecular processes limiting activity in solid state bioremediation reactors is possible, provided a multi- and interdisciplinary approach to the problem is adopted. Such models will not only increase understanding at a fundamental level, but will also give a sensible basis for process optimisation.

In general, models tend to emphasise empirical formulae for microbial biomass that ignore both physiological states and, hence, the metabolic capacities of the process microbes; they employ kinetic expressions selected on the basis of their ease of handling, in mathematical terms rather than on their empirical usefulness, as they are developed under very particular and specialised conditions that ignore critical process variables. For example, in recent years, various conceptual models have been developed for the design of activated sludge treatment plants. They seek to describe how plants will function after their construction. Because of the way they are formulated, however, virtually all ignore important fundamental microbiological concepts.

A similar limitation exists for mixed culture mediated biodegradation, in that the theory describing it considers a population of microbial cells to be homogeneous with respect to its metabolic properties. The simplified models are inaccurate and not completely predictive because they consider both metabolites and enzymes to be freely diffusible, they do not allow incorporation of long-term changes due to induction or repression, do not address the absolute magnitudes of either fluxes or metabolite concentrations, and are largely restricted to steady state systems.

What is clear is that a better understanding of biodegradation kinetics will undoubtedly enhance possibilities for more effective and efficient biotreatment. In this respect, an alternative formalism to metabolic control theory that is receiving increased attention is a thermodynamic model for microbial growth which is an assembly of independent metabolic compartments into which relevant biochemical information can be inserted. This model will respect both the laws of thermodynamics and kinetic principles.

6. Analysis and measurement

The rate of influx of new molecules into the environment far outstrips the development of analytical methods. There is therefore a need for wide-ranging technologies and devices capable of monitoring general levels of pollution as well as individual chemicals. Routine, real-time identification and quantitative analysis of pathogens are also required. In spite of the successful demonstration of the applicability of immunoassay methods in the scientific literature over the past ten to fifteen years, their potential has not been fully realised and implemented. The reasons are many. Part of the problem arises from misunderstanding and skepticism on the part of analytical chemists working in the field, and another part arises from the lack of studies conducted to provide thorough quantitative comparisons with established methods. A full understanding of the advantages and limitations of immunoassay methods is essential to applying the technology in the situations where they offer the most promise. Immunochemists need to make others working in environmental chemistry more fully aware of the versatility and general applicability of immunoassay methods.

Measurement is key to monitoring of process parameters is measurement. Nowhere is this more difficult than in the field (Case Study 13) and consequently standardised sampling methods need to be developed. It might be noted that measurement of the disappearance of a pollutant is not sufficient since the original material may be converted into a metabolite with high persistence or toxicity, often absorbed to the soil matrix or complexed by organic soil substances. It is also necessary to have methods for determining rates at which *natural* processes biodegrade contaminants. If these are rapid enough, it may be possible to show that the best remediation is to leave the system alone. In the reactor, on-line biomonitoring devices capable of quantifying incoming load and monitoring for potential toxic shock, and transferring information to the control operating systems, are required.

Case Study 13:   Leaking fuel from an underground storage tank, Traverse City, Michigan – United States

An aquifer was contaminated with JP-4 fuel from a spill that occurred in 1969 when a flange broke on an underground storage tank containing aviation fuel at a Coast Guard Air Station. Approximately 38 000 l of aviation fuel were spilled. The major groundwater contaminants were benzene, toluene, and xylene isomers. The subsurface materials at the site were highly permeable sand and gravel, making the successful application of *in situ* bioremediation likely.

Two *in situ* bioremediation field demonstrations were performed. Oxygen (supplied by hydrogen peroxide) was used as the electron acceptor in one, and nitrate in the other. Mathematical modelling of contaminant transport and biodegradation was used to aid in the selection of system parameters, and tracer studies were used to test the modelling results and evaluate system performance.

In both demonstrations, infiltration of water into the system was accomplished using infiltration galleries. Injection wells provide the most direct method for infiltration into the saturated zone, but give poor delivery of additives to the unsaturated zone, are prone to clogging, and have higher installation costs than infiltration galleries and surface application.

An important distinction between the two demonstrations is the choice of electron acceptor. Both electron acceptors should allow transformation of the aromatic compounds in the JP-4, and they did. However, although degradation of the aliphatic fraction of the JP-4 occurred with oxygen as the electron acceptor, little of this fraction was removed with nitrate.

In the oxygen-mediated demonstration, the concentration of BTEX initially decreased and then increased, as hydrocarbons previously trapped in the unsaturated zone dissolved into the groundwater as the water table rose. Oxygen breakthrough occurred at about 65 days, after which inorganic nutrients and BTEX levels decreased, while microbial numbers increased, indicating that biodegradation was occurring. After eight months, BTEX remediation had occurred up to 17m from the injection point.

Source: Environmental Protection Agency, United States.

7. Reaction kinetics and growth limiting factors

In all but the most heavily polluted environments, the microbes present have to cope with the presence and low availability of a multiplicity of substrates and nutrients. In most real environments it is very unlikely that a single compound will be growth-limiting. In recent years, considerable interest has developed in dual carbon energy substrate-limited growth and in dual electron acceptor-mediated growth, but the question of growth limitation by *multiple* substrates and nutrients simultaneously, has largely escaped investigation. In this area particularly, site-specific investigations, using a multi-disciplinary approach, are essential, as is practical research carried out at the location of contamination.

A most important aspect of substrate and nutrient availability is that the physiological potential of microbes depends primarily on their ability to synthesise enzymes; this in turn is a function of the availability of the nitrogen source. This has obvious implications when the biodegradation of carbonaceous pollutants occurs under nitrogen deficiency.

A further important question concerning substrate- and nutrient-limited growth and biodegradation is how appropriate are kinetics relating to the saturation of an enzyme when only very low substrate or nutrient concentrations exist. Adequate understanding of whatever kinetics are appropriate is critical in estimating residual concentrations and, hence, the effectiveness of bioremediation. It is also most important in complex systems to differentiate between kinetic limitations and diffusional limitations and to understand the nature of the release of adsorbed substrate and solid substrate solubilisation.

8. Process design and optimisation

Knowledge concerning both the dynamic response of cultures and the manifestations of unsteady state growth is meagre. More than 30 years ago some most illuminating investigations on phenotypic variability were published which showed the dramatic effect of growth rate on both bacterial cell size and on the composition of growing bacterial cells. It was also reported that when bacterial cells were transferred from an environment permitting high growth rates to one permitting much reduced rates, a distinct period of equilibration was required before bacterial cell size and composition corresponded to those typical for the bacterium in question under the new conditions and appropriate growth proceeded. What is most remarkable is that, even now, response data generated under transient state operating conditions are not widely available, so that the need for biochemical (bioprocess) design information cannot be adequately satisfied, and the design still has to rely on steady state data which often does not apply in continuously changing culture conditions.

New microbial processes are necessary, including co-metabolic as well as anaerobic, aerobic, and microaerophilic transformations. Compounds that are only co-metabolised represent a special category of concern because many of the current approaches to bioremediation are not relevant to such chemicals.

In order to maintain their metabolic activities, micro-organisms continuously need to transfer nutrients and their waste products between the external environment and the cell. During conventional aerobic processes, microbial demand for substrates other than oxygen is usually met without difficulty, as these are supplied in excess in the growth medium. Consequently, in many low-viscosity systems the supply of oxygen is the rate-limiting factor governing the metabolic activities of the cell. This is even more the case in high viscosity liquids.

In conventional activated sludge, the incoming air serves as a buoyant medium to agitate the mass and mechanical mixers are rarely used. However, the major interfering substance for oxygen transfer is nitrogen gas in the air bubbles, hence the use of purified oxygen rather than compressed air for high-rate reactions in aqueous media. Formation of extremely small bubbles and provision of a high degree of turbulence are essential to reduce the length of the diffusion path for oxygen molecules, since oxygen has very a low solubility in water, particularly at elevated temperatures. Technologies for the bubble-less transfer of oxygen into the reaction medium are being developed.

9. Bioavailability

Many compounds or pollutants that are quickly biodegraded under laboratory conditions persist in the environment. Although the pollutants may be destroyed by microbiological means, environmental factors and their physical or chemical state prevent their rapid degradation in the field. The location of the pollutant molecule, in the vapour phase, water or organic film or adsorbed onto the solid matrix may be of critical importance (see Figure 3). This constraint on bioremediation represents a major limitation to the widespread use of many biological technologies. The reasons for reduced availability of substances and ways to bring about the release and biological destruction of pollutants must be understood. While some substances sorbed on activated carbon appear to be more available to bacteria than when they are in solution, spaces between mineral layers are frequently small enough to admit organic molecules while preventing access for microbes or even their extracellular enzymes.

It has been found that microbial clean-up of soils or solids can proceed only to a certain minimum concentration, below which the contaminant does not come into contact with the micro-organisms. An understanding of the movement of micro-organisms in soil and on the surface of artificial media is essential to improving contact with the pollutant.

10. Recommendations for future R&D

The following proposals for discrete R&D activities are the recommendations of the Ad Hoc Group, as discussed during the preparation of this report. No prioritisation within the subdivisions was made. The advice given by an international panel of scientists specialising in the field of environmental biotechnology and elaborated during a workshop held in May 1993 in the United Kingdom was carefully considered and taken into account as appropriate. The full text of the paper produced at that workshop and the list of those who attended is included in the the Annex of Part I.

Organisms:
- Organisms and mixed cultures should be screened and selected for specific functional capacities (*e.g.* in waste treatment systems, specific defined organisms could replace current sludges with undefined populations of organisms);
- Organisms and their interactions in biocommunities, and communities in the environment should be studied;
- Organisms with novel and/or improved functional phenotypes should be developed either by recombinant DNA methods or through the selection of evolutionary mutants;
- Inocula with known functionality should be developed.

Bioprocesses:
- Processes are required based on the use of combined biological and physico-chemical techniques, *e.g.* membranes, in order to develop new bioreactor designs;
- Extensive and intensive processes are needed for the treatment of diffuse trace levels of pollutants;
- A better understanding is required of the nature and functioning of "black-box" processes, *inter alia,* carbon dioxide fixation, methane and hydrogen production, nitrogen and sulphur metabolism, production of biopolymers;
- Modification of environmental conditions, *e.g.* nutrients, acidity, etc., to enhance desired *in situ* biological conditions requires investigation;
- Processes occurring at surfaces, which underpin many biotechnological processes, need considerable research (*e.g.* kinetics, immobilisation, bioavailability, etc.);
- Model systems, *e.g.* microcosms, should be developed to predict operational performance and transfer techniques from the laboratory;
- Designs for technical processes are required for treating specific waste and extracting value-added products. Specific pollutants should be pretreated to avoid release;
- Processes aimed at prevention or treatment of waste should be directed at recirculation and added-value (clean water and air are value-added products);
- Investigations should be performed to stimulate a wider use of recovered products locally and also internationally; indeed environmental biotechnology should contribute to proper mass fluxes and global balances of nutrients and other materials by facilitating international commerce of the latter;
- Development of biological materials such as biopolymers, fuels and base chemicals to replace those produced from fossil fuels, is needed;
- Metals recovery by biological means should be improved, together with their subsequent recycling.

Ecology and ecosystems:
- A better understanding of how to restore and sustain impacted ecosystems is needed (for example, forests, lake and paddy field ecosystems);
- Mechanisms both of plant survival in arid conditions and of salt tolerance, need further research;

– Mapping of ''hidden'' microbiological diversity and development of new culturing techniques are needed to exploit natural resources better.

Measurement and monitoring:

– Measurement tools, procedures and protocols for site assessment should be improved;
– Sensitive but robust instrumentation and diagnostics need to be developed to identify both biological and chemical hazards, for instance, whole cell biosensors for water quality;
– Reliable test procedures must be developed to assess the fate of anthropogenic chemicals in full-scale treatment systems;
– New assays for specific pollutants are needed, including those based on immuno-technology;
– Better techniques need to be developed for monitoring mixed populations of micro-organisms and for tracking organisms within ecosystems;
– ''On-line'' toxicity tests should be developed to assess whether water streams or soils are contaminated or decontaminated;
– Operational clean-up criteria need to be formulated in order to determine how clean is clean the production of intermediates of degradation, levels of metal ions, etc.

Chapter 4

Future Applications of Biotechnology

This chapter contains some examples, rather arbitrarily chosen, of how biotechnology may improve the prevention, detection and treatment of pollution in the near to medium future. Some of these techniques are at pilot-plant stage and may well be applied commercially within this decade. Others are for the longer term and are just exciting possibilities at present.

Of particular interest is the possible use of biotechnology to resolve some of the global environmental problems that beset us, such as the global warming resulting from CO_2 production and the increasing desertification of the planet's surface.

1. *In situ* bioremediation

Consortia of organisms

Bioremediation processes have invariably depended on mixed cultures. In virtually no open unprotected environment, engineered or natural, have conditions either selected for or preserved monocultures of micro-organisms. The steady-state, single organism condition in the laboratory is very far removed from the irregular, multi-organism, multi-phase set of circumstances that represent the field condition. As far as bioremediation is concerned, the consortia of micro-organisms that are likely to generate most interest are those that achieve biodegradation of otherwise recalcitrant materials. The best known consortia are the various methanotrophic consortia in which the non-specific activity of the methane mono-oxygenase is exploited. Similarly, the ammonia mono-oxygenase of *Nitrosomonas* spp. in chemoautotrophic-nitrifying consortia also offers potential, especially if growth rates and bacterial concentrations of *Nitrosomonas* spp. can be markedly enhanced. Research into the interaction of selected species in microcosms holds out the promise of higher-rate, more compact bioreactors. The future will bring increased use of consortia adapted for the bioremediation of specific target pollutants.

Genetically modified organisms

Recombinant DNA technology is being used to develop superior strains of micro-organisms to speed up degradation and widen the range of easily degradable compounds. One such organism has already been patented. Pure cultures of selected or engineered micro-organisms will play a major role in the future to abate pollution. Industrial effluents are particularly suitable for directed degradation because the starting material is reasonably well-defined. Contained use in bioreactors of GMOs selected for particular recalcitrant molecules will be the first such applications. It is likely that GMOs will ultimately be used *in situ*: laboratory research and novel detection capabilities are going hand in hand to make this possible with little risk, although considerable R&D in microbial ecology will be required.

Regulatory requirements for the release of genetically engineered organisms may, in some nations or some instances, militate against their early use in bioremediation. Additionally, scientific and engineering uncertainties and public concern may result in a slow start to the *in situ* use of GMOs. Since it will be possible to isolate a much wider variety of different organisms, capable of degrading recalcitrant compounds, it is possible that the selection of naturally occurring organisms and the use of GMOs will go hand in hand.

2. Real-time monitoring and delivery systems

Among the technical advances, an important development will be the use of biosensors for the remote detection of pollution incidents. Their chemical specificity and low power requirements make them ideal instruments for unattended monitoring in the environment. Made to respond to a variety of potential pollutants, biosensors will be constructed to detect low concentrations which can be reported by radio or recorded on a continuous basis. A warning can be sounded in the event of a discharge and a record provided of the magnitude of the event and the effectiveness of remedial measures; conceivably, corrective action might, in many instances, be instituted directly by the sensor without waiting for human intervention.

3. Bioprocessing

Advanced bioreactors

The search for greater efficiency and better economy in the treatment of municipal and industrial waste, and the promulgation of stricter pollution control legislation continue to stimulate the development of advanced bioreactors. In particular, control of sea-dumping of sewage is driving the development of high- throughput, relatively small footprint reactors for siting at the water's edge.

A better understanding of the ecology of mixed cultures, the physiology of micro-organisms, and hence control mechanisms will be the basis for this progress. In the future, there will be a switch from empirical processes to developments based on a multi-disciplinary approach incorporating genetics, microbiology, engineering and mathematical modelling. In the wider field of bioremediation there will appear both high rate (< hours) and very low rate (> years) treatment systems.

Novel separation techniques

Industrial effluents, in addition to their organic content, may be highly alkaline or acidic and have high salt concentrations, both sets of conditions being hostile to micro-organism activity. Rather than neutralise the effluent and precipitate the salts before treatment, it may be possible to construct a bioreactor in which the organisms are separated from the incoming effluent by membranes which permit only the organic pollutants to pass. A number of experimental reactors using elastomeric membranes have been tried. Microfibril membranes, used as supports for biofilms, can combine separation and degradation of pollutants into one operation.

4. Emerging global threats

Photosynthesis and carbon dioxide fixation

The atmospheric concentration of carbon dioxide, a major cause of the greenhouse effect, is increasing with the consumption of fossil fuels and the felling of the world's forests. By improving the mechanism of photosynthesis by biotechnological means and using solar energy more effectively, large volumes of CO_2 exhausted from industrial plants might be fixed and a variety of useful substances recovered. Photosynthetic micro-organisms, such as microalgae, have a higher photosynthetic ability than large plants. However, even they only use a small proportion of solar radiation. One way of addressing the problem of increasing CO_2 may thus be to use advanced biotechnology to significantly improve the photosynthetic ability of plants and the CO_2-fixing capacity of corals and shellfish.

On the negative side, it has been reported that UV-B radiation (wave length 280-320nm) inhibits photosynthesis in microalgae at intensities which do not affect photosynthesis of terrestrial C3 plants. The breakdown of the ozone layer will cause an increase in UV-B radiation at the earth's surface, which may in turn result in a decrease in ocean productivity.

Prevention of desertification

Present deserts appeared on earth some 10 000 years ago, and the area affected may be expanding by as much as 60 000 sq km a year. Currently, 4.5 million sq km or 35 per cent of the global land area are threatened by desertification. The applications of biotechnology for preventing this phenomenon may be divided into two groups, short- and medium-term technological developments and long-term basic studies. The main short- and medium-term issues include water retention and the prevention of salt damage. A research group in Japan has developed a new "super-bioabsorbent" which can absorb and hold water at more than a thousand times its own weight. A long-term priority is the breeding of plants, such as xerophytes and salt-proof plants, compatible with desert conditions, using gene recombination and cell fusion techniques.

5. Production of clean fuels

Hydrogen production

Hydrogen is required for oil refining processes and the effective utilisation of carbon dioxide. It is also considered a possible replacement automotive fuel. Technologies are required to produce hydrogen effectively without consuming fossil fuels. With the aim of developing the microbial production of hydrogen, photosynthetic micro-organisms are being screened for high yield of hydrogen and for massive cultivation to maximise hydrogen production ability. Studies are also being carried out on effective separation and refinement technologies for hydrogen, large-scale hydrogen production systems, and the operation of pilot plants.

Establishment of these technologies will be expected not only to suppress the CO_2 from current hydrogen production but also to contribute to the effective utilisation of CO_2 from other processes.

Biodesulphurisation of oil and coal

Energy sources of fossil origin contain organic sulphur (*i.e.* carbon-bonded). In the case of coal, inorganic sulphur, in the form of sulphides (usually pyrites) is frequently trapped within the matrix. The removal of sulphur from fossil fuels is an important issue in relation to global environmental problems, particularly acid rain. Although current oil desulphurisation technologies are efficient, high temperatures and pressures are required to achieve further removal. New requirements to decrease sulphur in light oil and kerosene from 0.5 per cent to 0.05 per cent in five years are likely to be enforced. Some organic sulphur compounds with a thiophene skeleton are not removed by current treatments.

It is therefore important to study the possibility of biodesulphurisation as an environmentally friendly option. Several micro-organisms, including several species of *Thiobacillus*, are able to remove pyritic sulphur from coal by the oxidative metabolism of insoluble sulphides to water soluble sulphates. Many different microbes are being evaluated for the removal of organic sulphur, which must be carried out under anaerobic conditions to prevent the risk of explosion. The critical step for the specific removal of organic sulphur is the cleavage of C-S bonds, particularly in dibenzothiophene rings, to allow solubilisation and removal of the sulphur atom while leaving intact the rest of the carbon matrix. This reaction seems to be very infrequent in nature and only a few micro-organisms have so far been reported to effect it on model compounds. At the Institute of Gas Technology in the United States a micro-organism to split carbon-sulphur bonds was isolated (under aerobic conditions), and in Japan a group from the Fermentation Research Institute isolated a micro-organism under microaerobic conditions which assimilated and degraded alkyldibenzthiophene.

Even with such strains, the S-containing substrates may not be readily bioavailable and prospects of a successful process are dependent on solving problems such as substrate contact with membrane-bound enzymes. Work is progressing to engineer organisms to increase enzyme activity and tolerance to acid (in the case of coal) so that a combined process for removal of inorganic and organic sulphur can be developed.

Chapter 5

Sources of Information

This chapter is intended to direct the reader towards further sources of information. The references included in the brief bibliography here should provide some expansion of the information in the report while the databases referred to offer a much wider range. A much more complete list of scientific references is included in a separate document available on request. A descriptive list of many of the currently available databases is also available in a separate document, again available on request, in which each database is described and availability and contact information is given.

Bibliography

American Academy of Microbiology, *Scientific Foundations of Bioremediation – Current Status and Future Needs*, (1992), Washington DC.

Atlas, R.M. and Bartha, R. (1992), *Hydrocarbon Biodegradation and Oil Spill Bioremediation, Advances in Microbial Ecology*.

Gallagher, J.R., Rittman, B.E., Valrichi, A.J., Seagren, E., Ray. C. and Wrenn, B. (1992). *A Critical Review of In situ Bioremediation*, Gas Research Institute, Chicago.

Hall, D.O. and Krishna Rao, K. (1989). "Immobilised photosynthetic membranes and cells for the production of fuels and chemicals". *Chimicaloggi*, March, 42.

Halvorson, H.O., Pramer, D. and Rogul, M. (eds) (1985), *Engineered Organisms in the Environment: Scientific Issues*, American Society for Microbiology, Washington DC.

Hance, B.J., Chess, C. and Saidman, P.M. (1990), *Industry Risk Communication Manual*, Lewis Publishers, CRC Press, Boca Raton, Florida.

Hinchee, R.E. and Olfenbuttel, R.F. (eds) (1991*a*), *In situ Bioreclamation: Applications and Investigations for Hydrocarbon and Contaminated Site Remediation*, Butterworth-Heinemann, Boston.

Hinchee, R.E. and Olfenbuttel, R.F. (eds) (1991*b*), *On site Bioreclamation: Processes for Xenobiotic and Hydrocarbon Treatment*, Butterworth-Heinemann, Boston.

Lens, P. and Verstraete, W. (1992), "Aerobic and Anaerobic Treatment of Municipal Waste Water", in: Villa, T.G. and Abalde, J. (eds), *Profiles on Biotechnology*, Universidade di Santiago de Compostela, pp. 333-356.

Levin, M.A., Seidler, R.J. and Rogul, M. (eds) (1992), *Microbial Ecology, Principles, Methods and Applications*, McGraw-Hill, New York.

Liessens, J., Germonpre, R., Beernaert, S. and Verstraete W. (1993), "Removing Nitrate with a Methylotrophic Fluidized Bed: Technology and Operating Performance", *Journal AWWA 85*, pp. 144-154.

Mergaert, K. and Verstraete, W. (1992), "Applicability and trends of anaerobic pre-treatment of municipal waste water", *Water Res.*, 26, pp. 1025-1033.

Office of Technology Assessment, (1991), *Bioremediation for Marine Oil Spills*, US Congress, Washington, DC.

Omenn, G.S. (ed) (1985), *Environmental Biotechnology, Reducing Risks from Environmental Chemicals through Biotechnology*, Plenum Press, New York.

Porta, A. (1991), "A Review of European Bioreclamation Practice", in: Hinchee, R.E and Olfenbuttel, R.F. (eds), *In Situ Bioreclamation* , Butterworth-Heinemann, Boston, pp. 1-13.

Pritchard, P.H. and Costa, C.F. (1991), "EPA's Alaska Oil Spill Bioremediation Project", *Environmental Science and Technology*, Vol. 25: pp. 372-79.

Schultz, J.S. (1991), "Biosensors", *Scientific American*, Vol. 265 No. 2: pp. 64-69.

Verstraete, W. and Top, E. (1992), "Holistic Environmental Biotechnology", in: J. Fry, G. Gadd, R. Herbert, C. Jones and I. Watson-Craik (eds), *Microbial Control of Pollution*, Cambridge University Press, Cambridge, pp. 1-18.

United States Environmental Protection Agency (1992), *Bioremediation of Hazardous Wastes by Biosystems*, Technology Development Program, EPA/600/R-92/126.

Databases

The information being so rapidly generated and gathered in the field of biotechnology can be readily accessed via several media, including print, online and CD-ROM databases. These databases deal with all the various aspects of biotechnology, including theoretical, commercial, pharmaceutical, medical, and environmental. Sources include scholarly journals, trade publications, government and business reports, and the popular press. The databases are either bibliographic, that is, they are lists of citations to journal articles, books and other publications in the literature of biotechnology, or they function as storehouses for the actual data generated in laboratories and in the field.

Database examples:

IRRO – Information Resource on the Release of Organisms into the Environment.

IRRO is not a database but rather an electronic network that will provide centralised access to existing data sources in different regions of the world. The network uses modern telecommunications systems and sets up gateways and interfaces. It provides a single contact point for all those studying the release of organisms into the environment. The activity was initiated by UNEP.

It is intended that information on both non-modified and genetically modified organisms should be available and that macrobiota as well as microorganisms should be included.

The focal point for IRRO is MSDN (Microbial Strain Data Network), Huntingdon, United Kingdom.

AGRICOLA – A database of the US Department of Agriculture, National Agricultural Library, Beltsville, Maryland.

A search made in the "Quick Bibliography" Series (QB 91-106) revealed 160 citations on the topic of biotechnology and bioremediation which had been entered in the AGRICOLA database between January 1979 and March 1991. These include books, conference reports, journal articles and theses from around the world.

Copies of bibliographies in this series may be obtained from the National Agricultural Library. Requests for articles, particularly from overseas, should be submitted through major university, national or provincial institutions.

ATTIC – Alternative Treatment Technology Information Center.

ATTIC is a comprehensive information retrieval system containing V7v4 data on alternative treatment technologies for hazardous wastes. The ATTIC system is a collection of hazardous waste databases that are accessed through a bulletin board. The bulletin board includes features such as news items, bulletins, and special interest conferences such as the Bioremediation Special Interest Group. It also features a message board that enables users to "chat" and request advice from other users, as well as pose questions to the ATTIC System Operator. ATTIC users can access any of four databases:

- the main ATTIC Database, which contains approximately 2 000 records dealing with alternative and innovative technologies for hazardous waste treatment;
- the Water Treatability Database, which provides data on characteristics and treatment of a wide variety of contaminants;
- a Technical Assistance Directory, which identifies experts on a given technology or type of contaminant;
- a calendar of upcoming relevant conferences and events.

In addition to the above listed databases, which can be accessed online, other databases may be accessed through the ATTIC System Operator. These databases include the Cost of Remedial Action (CORA) Model, the RSKERL Soil Transport and Fate Database, and the Geophysics Advisor Expert system. The System Operator is also available to assist users with any problems, answer questions, or do searches. Hardcopy of documents can also be ordered through the System Operator.

At the present time, an online database to identify potential sources possessing qualifications to perform specific types of treatability studies is being developed. This online database will supplement the existing hardcopy version of the Inventory of Treatability Study Vendors, which has previously been made available to the EPA Regional Offices and EPA contractors.

ATTIC can be directly accessed by users with a personal computer and a modem. New users can register themselves and assign their own password by calling (301) 670-3808 through their modem.

MINE – Microbial Information Network Europe.

This network contains data on more than 100 000 micro-organisms of the main culture collections from Belgium, Finland, France, Germany, Greece, Italy, the Netherlands, Portugal, Spain, Sweden and the United Kingdom and is the most comprehensive microbial strain database existing at present.

In order to be able to integrate data from different collections, all MINE collections adhere to a common format for the storage and retrieval of strain data. In this format some 120 fields have been defined and include: species name, strain environment and history, biological interactions, properties (cytology, biomolecular data), genotype and genetics, practical applications (including the field of environmental biotechnology), pathogenicity.

Appendix 1

Other Biotechnological Activities with Impact on the Environment

Agrofood biotechnology

The last half century has seen the introduction of more intensive agricultural practices, often with a greatly increased use of agrochemicals and water. In OECD countries, this is now recognised as the source not only of surplus agricultural production but also of considerable environmental pollution.

During the last ten years, a rapidly growing number of biotechnologies have been developed to modify the production of plants and animals, as well as their conversion into food and non-food products, and many of them have environmental relevance. The main report cannot give a detailed review of these new biotechnological options, and the reader is referred to the large and growing literature, particularly a recent OECD report entitled "Biotechnology, Agriculture and Food" (Paris, 1992). However, it is useful to illustrate the interface between agro-food and environmental biotechnology at various levels and in various sectors, since the former can be seen as part of environmental biotechnology in the wider sense.

Most agricultural biotechnologies are targeted towards increasing productivity. These increases can reflect either improved yields per unit of input or decreasing inputs and costs per unit of output or both. In countries where the problem is food overproduction rather than shortage, the new thrust will be towards reducing costs and inputs, meaning less land, water and agrochemicals for the same amount of food – a desirable environmental trend.

New biotechnology can affect every stage of plant life – breeding, growth, harvesting and residue treatment – and at each stage there could be a consequent benefit for the environment in the form of more efficient, less resource-consuming, less polluting agricultural practices. Biotechnological developments are permitting the more rapid breeding of higher-yielding plants having a higher proportion of harvestable material. Fundamental breakthroughs may come from new genetic modification methods that aim at increasing plant resistance to virus and other diseases, as well as to drought, salt, cold, and heat. This will enlarge the land resources available for crop production. Another priority goal of modern plant genetics is the replacement of nitrogen fertilisers, another major source of pollution, with nitrogen fixation within the plant.

Animal biotechnology could greatly benefit the environment in regions where waste from intensive animal rearing is leading to serious soil and water pollution. In such regions it is becoming vital to produce the required quantities of meat and milk from fewer animals. The main advances are coming from new animal health products, improvement in animal growth and lactation (*e.g.* the use of bovine somatotropin, BST), embryo multiplication, improved animal feeds (and feed conversion), and genetic modification to introduce new traits.

Some biotechnologies are already in use and are likely to have a beneficial effect on the environment. Many more are technically ready but their large-scale use is not yet permitted due to safety or public opinion concerns. Others are still in the development stage.

Conversion of agricultural raw materials into food and non-food products, including wood, pulp and paper, and leather, has traditionally contributed large amounts of industrial waste. Improvements in production processes in these industries through biotechnology, for example, the replacement of harsh leather tanning chemicals by enzymes, with the aim of reducing and ultimately preventing the generation or release of waste, are increasingly possible. Such improvements can involve the conversion of wastes into useful and valuable products: almost 10 per cent of the value of the wheat crop is now derived from the conversion of straw into starch and other industrial products with the help of new enzyme technologies.

In the long term, the most important contribution to the prevention of environmental damage will come through substitution of traditional, intensive practices of crop and livestock production by newer ones based on biotechnology. However, long before biotechnology will help the agro-food sector to reduce or prevent waste, it will have to look at the waste streams generated by the sector. These, which include high BOD effluent from food processing plants, seepage of pesticides and nitrogen fertilisers into rivers and lakes, slurries of animal wastes in excess of what can be spread on the land and mal-odours from intensive animal rearing, can be approached by the technologies described in detail elsewhere in the report.

Biosensors, based on the interaction between target chemicals and biological systems, can also aid in the detection and monitoring of effluents from agricultural production and are being developed, as described in the report, for the detection of residues in water, soil and food materials. Rapid biochemical tests for the presence of crop disease organisms make these diseases detectable at an earlier stage, allowing farmers to choose appropriate chemicals and to use them at lower application rates. These chemicals, which are a source of pollution, are gradually being complemented if not replaced by biological control methods.

Biomining

The capacity of some acidophilic bacteria to oxidise mineral sulphates can be utilised for the release of metals from ores, concentrates or waste materials. The technology is variously referred to as bacterial mineral leaching, biomining, or biohydrometallurgy. The reverse of this process – bioabsorption or precipitation of metals – may fall under the heading of bioremediation and legitimately form part of the main report.

There can be technical and economic advantages in utilising bacteria for mineral sulphide oxidation. The biotechnology can be increasingly seen to have appeal as a "clean technology" when compared with some conventional processes, for example, the avoidance of emissions associated with smelting of sulphides. However, it should be noted that the predominantly acid, potentially toxic metal-laden liquids of biomining would give rise to pollution if not contained, recycled or treated.

The impact of biomining will always be restricted to particular combinations of minerals and circumstances. Nevertheless there are successful, commercial mineral-processing operations involving bacteria. These are basically of three types. First, there are dump- or heap-leaching processes, in which the bacterial activity causes the release of target metals, principally copper and uranium, into percolating acid water. Second, there is underground or *in situ* leaching of uranium. Being largely underground, this operation has greatly reduced environmental damage normally associated with uranium mining and surface tailings deposition. Third, bioreactors can be used for processing high-value (gold-bearing) concentrates.

Dump, heap and *in situ* operations rely on natural populations of indigenous organisms that proliferate whenever and wherever mineral sulphides are exposed to water and air. Naturally occurring populations of bacteria exist at mine sites, coal spoil sites where the coal contained pyrites and in acidic hot springs and geothermal areas. It is likely that optimisation of conditions (*e.g.* pH and addition of nutrients) would ensure denser and more active populations than those developing naturally.

The microflora found at these sites often contain the same strains of identical or very closely related bacteria despite considerable geographical isolation. For example, the same strains of *Thiobacillus ferrooxidans* and of moderate thermophiles have been isolated from copper leaching dumps in the United States, Icelandic hot springs and their run-off, coal spoils in England, and mine sites in Russia. There are, however, exceptions, and the species present are inevitably very mixed and it is likely that many of the types in communities have yet to be described. The mining industry is not generally concerned about the nature or types of bacteria that degrade minerals as long as they perform reliably, reproducibly and in a "user-friendly" fashion, without the need for regular attention that would require microbiological expertise.

It is possible that improvements in activity through genetic manipulation could facilitate the recycling of dissolved ferrous to ferric iron and be exploited in waste treatment. It is difficult, however, to envisage genetic manipulation resulting in improvements which would enhance waste treatment as significantly as could optimisation at the bacterial growth conditions and process engineering. It is also not clear that there would be any advantage in attempting to improve the fundamental capacity of mineral-oxidising bacteria with regard to the rate of mineral sulphide oxidation, since it is very unlikely that the organisms will be operating near to their maximum potential in industrial systems.

The most likely problems to be overcome might involve potential inhibition of the bacteria by, for example, toxic metals, organic reagents or cyanides. However, the exposure of bacteria to gradually increasing concentrations of toxic metal ions often leads to adaptation involving selection of mutations in the natural population. In one case, the organisms apparently developed sufficient tolerance to arsenic to make genetic manipulation unnecessary.

There is likely to be an increasing application of bacteria for reducing pollution in the mining industry. For example, biological treatment processes are being used to remove cyanide and metals from mine water (Case Study 14). Micro-organisms have been used to separate heavy metals from solutions in order to detoxify them and to recover precious metals from industrial waste. It is notable that the National Institute of Standards and Technology in Japan is currently investigating the use of metal-metabolising micro-organisms for resource recovery, bioremediation, and coal cleaning.

Biosorption processes have been tested as clean-up systems, and any industrial process would probably use dead biomass as a substitute for ion-exchange resins. However, some work has shown that living (genetically engineered) strains could out-perform natural enrichment cultures in removing mercury from waste water. These techniques are considered more fully in the main report.

Case Study 14:   Clean-up of drainage water from an abandoned mine in Japan

After the closure of the Matsuo sulphur and iron sulphide mine, acid mine drainage water continued to contaminate the Kitakami river. The local authorities petitioned the Government for a solution and the decision was taken to build a neutralisation plant. Construction started in 1977 and was completed in 1981 at a cost of about ¥6.1 billion.

The drainage water is very acid (pH 2) and contains iron and arsenic. Ferrous iron in the water is oxidised by bacterial action to insoluble ferric sulphate using a concentrated culture of *thiobacillus ferrooxidans* recycled from the sludge settling tank. Subsequently, the water is neutralised using calcium carbonate, and a further sludge of ferric hydroxide is generated. The combined sludges are coagulated and settled and are collected in a 2 million m^3 capacity dam; 9 million m^3 of mine water is treated a year and the plant runs at a rate of 15-20 m^3 per minute. There is an increase in pH from 2 to 4, iron is reduced from 400 to 2 mg/l, and arsenic from 70 to 0.1 mg/l.

In addition to cleaning up the drainage water, the authorities are undertaking civil engineering work to prevent further mine-related pollution.

Source: Metal Mining Agency of Japan.

Appendix 2

Microbial Consortia

To date, bioremediation processes have depended exclusively on mixed cultures. In virtually no open unprotected environment, engineered or natural, have conditions either selected for or preserved monocultures of micro-organisms. These environments are characterised by a diversity of microbial strains and by changing selective pressures which induce successional changes in community composition based on self-regulatory interactive mechanisms between community members. It is important to note that current applications of bioreactors for the remediation of waste materials also depend on the use of a variety of mixed cultures.

Within the complex and diverse populations found in nature there exists a variety of interactions including:

- neutralism: no interaction;
- mutualism: each member of the population benefits from the other;
- competition: micro-organisms compete for available nutrients and space;
- parasitism: one organism benefits at the expense of another;
- amensalism: one adversely changes the environment for another;
- synergism: a non-obligatory interaction between two or more populations wherein both populations benefit; and
- predation: an interaction in which one organism ingests another.

One might expect that when a mixed culture is enriched for strains that utilise a specific carbon energy substrate, the result would be either an essentially pure monoculture or a mixture of strains that are all capable of utilising the specific carbon energy substrate, depending on the enrichment procedure employed. Although either possibility can be shown to occur in artificially assembled mixed cultures, both possibilities are rare occurrences in the case of enrichment of mixed cultures from natural environments. The usual outcome is a microbial consortium comprising a single primary-substrate utilising strain growing in association with between two and five ancillary non-primary-substrate utilising strains (Figure 13) that optimise the growth environment for the primary strain, but collectively represent only a minor fraction of the consortium.

In cultures of the type that mediate biodegradative activities in complex polluted environments, primary interactions must be expected to occur not between individual strains, but between various specific primary-substrate utilising consortia. However, even here, under essentially stable operating conditions, what one might observe is an apparent absence of interaction. One of the most interesting aspects of interacting consortia is whether ancillary strain transfer between consortia occurs. If barriers do exist to ancillary strain transfer, this would explain some of the difficulties that are encountered in bioaugmentation.

Widely studied are the consortia involved in co-operative metabolism, in which a series of primary-substrate degrading strains, each possibly associated with ancillary non-primary- substrate utilising strains, are responsible for either a step or steps in the biodegradation of an original primary substrate. Such systems have been extensively discussed with respect to the degradation of various herbicides. In many respects the most extensively studied consortia with respect to co-operative metabolism are those involved in the anaerobic methanogenic biodegradation of sugars, carboxylic acids (other than acetate) and other soluble compounds to form methane. A feature of these latter systems is the existence of two alternative methane-producing steps, *i.e.*, from acetate, and by the reduction of carbon dioxide with hydrogen. A further example of a consortium involving co-operative metabolism concerns the chemoautotrophic nitrifying bacteria, where *Nitrosomonas* spp. oxidise ammonia to nitrite and *Nitrobacter* spp. oxidise the nitrite, so produced, to nitrate.

Some of the best examples of enrichment cultures, comprising several specific consortia, involve the bio-oxidation of synthetic waste waters containing simple organic compounds. In the case of the simultaneous bio-oxidation of methanol, phenol, acetone and isopropanol, three separate consortia are responsible. One consortium oxidises methanol, the second oxidises phenol, while the third oxidises both acetone and isopropanol, but perhaps

Figure 13. A model of a three-membered consortium anaerobically converting ethanol in an effluent to methane and carbon dioxide

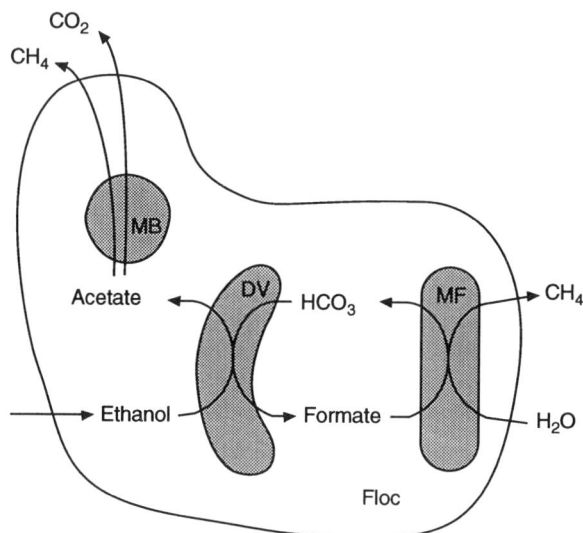

Notes:
DV = Desulfovibrio vulgaris.
MB = Methanosarcina barkeri.
MF = Methanobacaterium formicicum.
Source: R. Atlas, University of Louisville, United States.

most remarkable of all is the fact that the three consortia are sufficiently non-interactive so that a state of neutralism prevails between them.

Recently, it has been shown that for the simultaneous bio-oxidation of methanol, acetone, isopropanol and methylene chloride, superior rates were achieved when two primary-substrate utilising consortia, rather than a binary mixed culture comprising only the two primary-substrate utilising strains, were responsible.

If the degradation of organic pollutants primarily to biomass and carbon dioxide is likely in many cases to require a microbiological consortium for effective action, then perhaps constancy of the microbial population is neither an essential nor indeed an achievable objective: virtually any organism that can survive and grow by contributing to the breakdown of the pollutant(s) may participate. Consortia can be particularly valuable and cost-effective when the chemical nature of the pollutants is variable. A microbial population may then be able to adjust and adapt over time to the actual chemical composition of the material to be degraded. There is increasing evidence of horizontal gene flow with communities maintaining a high gene pool via plasmid-borne genes.

Waste Prevention and Cleaner Technologies

Waste prevention

Biotechnology can be used in four ways to prevent environmental damage:

a) **added-value processes** are those which convert a waste stream into useful products;

b) **end of pipe** processes in which the waste stream is purified by micro-organisms and the products are then released without harm into the environment;

c) **process innovation,** or the development of new biological processes (including the modification of existing ones) to diminish waste production;

d) **new biomaterials development,** or the manufacture of materials with reduced environmental impact.

Process options are to be preferred over end of pipe treatment, particularly if they have economic advantages. Potential examples include:

– the production of fatty acids, amino acids or solvents from carbohydrate wastes;
– the production of biomass for animal feed from waste streams of food processing plants;
– the production of yellow sulphur by microbial oxidation of sulphide-containing wastes;
– recovery of metals using micro-organisms (see below);
– production of methane from organic wastes.

An example of an added-value treatment is given by a strain of the thermophilic bacterium *Bacillus stearothermophilus* which can synthesise catechol from phenolic waste streams.

Traditional methods have a number of advantages and disadvantages. On the one hand they can be operated at low concentrations, and large waste volumes are effectively handled; they work at ambient temperature and pressure and they can be chemical-specific. On the other hand, they have low conversion rates, are easily disturbed and often have a long start-up time. A variety of disadvantages may also limit the production of added-value products. For example, the economics are generally worse than those of processes primarily designed to provide the same product. For example, a solvent may be produced from very dilute carbohydrate wastes (1-30g/l), using only relatively small waste streams (*e.g.* 1-30 0tonnes of carbohydrate/day). However, because of the small scale of this operation and the expensive downstream processing, costs are very high even when substrate costs are effectively "free".

If a new product is generated from a waste stream, then a long period of market development may be needed and product quality must be guaranteed despite the potentially variable feedstock from the waste stream. Experience shows that this market development can take from three to six years as, for example, in the production of yeast-vinasse and penicillin-biomass for feed additives by Gist Brocades. Sometimes a market cannot be found. Even for those processes which have relatively easy downstream processing, such as sulphur recovery, the small production volume and the product quality may easily result in unfavourable economics. These problems are all related to products which must be sold by the company generating the wastes. They do not exist if the product can be recycled in-house, thus diminishing the requirement for raw materials. The best example of such a process is the production of biogas which may be used to replace natural gas. It should be borne in mind that these processes rarely lead to complete purification of the waste. There will always be a need for post-process treatment to produce a stream which can be disposed of into the environment.

The decision to build on-site treatment facilities rather than dispose of the waste to municipal treatment depends on the availability of appropriate technology and relative cost. Air purification can, of course, only be carried out on-site.

Appropriate technology has to take into account a number of considerations in order to be applied at production plants. These include:

- treatment as close as possible to the source to avoid pre-treatment dilution;
- the limits of space available;
- the regulations on safety, hygiene and secondary emissions;
- the variability of waste water composition;
- continuously increasing effluent standards, so that lower concentrations must be achieved.

These requirements mean in practice that:

- reactors must be small in volume and area, in order to be placed close to the point of emission;
- reactors should be completely closed to avoid emissions of odour, aerosols and noise. This is especially critical if the treatment occurs in heavily populated areas or inside a large building;
- treatment systems must be designed specifically for each waste case;
- designs must be tailor-made, with maximum flexibility.

Every biotechnological process leads to microbial growth and the production of surplus sludge. Where this sludge is regarded as chemical waste in its own right, it may have to be incinerated (perhaps due to its concentration of heavy metals or undesirable chemicals). While there is a need for techniques to decrease this surplus-sludge production, they may make very little difference to the overall cost if the sludge has to be incinerated; most of such costs are capital rather than operating.

Cleaner technologies

Biology can be the basis for cleaner technology by eliminating a specific pollutant, either by replacing the chemical or by making its use unnecessary. The biological process can be implemented as a step in the actual production process. One example might be the use of a biological system to destroy excess solvents during process use. Such in-line processes are novel, however, and they have yet to be fully commercialised. Most discussion of cleaner technology revolves around the replacement of conventional physico-chemical processes by biological processes.

Conventional physical, chemical or thermal processes generate large amounts of waste and by- products through the use of:

- high temperatures to achieve high conversion rates (which may lead to a high rate of by-product formation, increase in temperature increases all reaction rates);
- extremes of pH to achieve acid/base catalysis (with the production of large quantities of salts when neutralising the reaction mixture);
- a wide variety of highly reactive chemicals to provide protecting chemical groups (such protection is needed to achieve selectivity of conversion);
- organic solvents in synthesis and purification.

In contrast, biotechnological processes generally occur at moderate temperatures (10-60 °C), moderate pH, and in aqueous systems. The applied enzymes or micro-organisms are highly selective, and the use of additional chemicals is minimal.

Such considerations would suggest that biotechnological processes should be highly advantageous. Nevertheless only a few production processes are actually in operation, for example those for the production of aspartame, 6-amino penicillanic acid (6-APA), p-OH-D-phenylglycine, and L-amino-acids. This situation is understandable: biotechnological processes are inherently expensive due to high capital and process development costs, which mean that biotechnology is currently only a valid option for high added-value products (cost price typically >$5-10/kg). A second aspect which hampers the introduction of a biotechnological process is the fact that it may represent only one unit operation in a whole process train of chemical and physical units, thus posing an integration problem.

Another consideration is that the high selectivity of the enzymes or micro-organisms, seen as an advantage elsewhere, may give rise to the requirement for a new enzyme for each new product. Because the development costs of a new enzyme can be high (of the order of $2 million to $10 million), the obstacles to such processes are equally high (compare this with the "cheap" route of high temperature and extreme pH to achieve a chemical conversion). Therefore, there is currently a trend to develop enzymes which have a broad substrate spectrum, a successful example being the kinetic resolution of various racemic amino-acid amides using the same enzyme (DSM-process).

A major problem with the use of enzymes has been their solubility in water, which makes them difficult to separate from substrate and products and, consequently, not suited to repeated use. These limitations can be overcome by immobilising the enzyme on a support medium. This may, in some instances, increase the stability of the enzyme, thereby increasing the efficiency of the process and decreasing both the raw materials consumed and the amount of waste produced. Additionally, novel protein engineering techniques are resulting in the production of modified enzymes able to function in the presence of organic solvents.

A process which should reach full development in the near future is the production of optically active compounds. The classical chemical processes are complex although advances are being made in homogeneous and stereo-specific catalysis. Widely available hydrolase enzymes are very suitable (acylases, dehalogenases, aminohydrolases, lipases, esterases) and such enzymes are even more attractive because they also catalyse reactions in a water-poor environment.

Biomass and biogas – an alternative energy source

Biomass is a lignocellulosic (woody, plant) material which can be a primary product or a waste product from both the agricultural and forest products industries. Perhaps its most attractive feature is that it is a product of photosynthesis and, as such, a renewable resource. The composition of biomass varies from source to source, but lignin, cellulose and hemicellulose are always present as major components. For conversion of biomass with maximum energy conservation, anaerobic digestion of the biomass by a complex consortium of bacteria and archaebacteria to produce methane is the most favoured, and extensively investigated, process, functioning at ambient pressures and at nearly neutral pH and converting biodegradable organic matter into methane and carbon dioxide (biogas) in the absence of oxygen. Depending on the ultimate use of biogas, both its significant carbon dioxide and water vapour content can be detrimental.

Essentially three basic types of biogas production exists: from landfill, from dedicated sources of biomass and as a by-product from anaerobic treatment processes for sewage sludge, domestic animal slurries and high-strength industrial waste streams in appropriately designed bioreactors (Figure 14). In the case of waste materials,

Figure 14. **An aerobic biogas plant**

Source: Krüger Bigadan, Denmark.

it should be remembered that the primary process objective of such systems is effective waste treatment rather than efficient biogas production. The formation of biogas represents an efficient method for recovering chemical energy from very wet organic waste. The chemical energy in sludge is just about enough to sustain combustion if the sludge is dewatered to 20-25 per cent. solids. This is close to the upper limit for filters or centrifuges. As biogas, the fuel automatically separates from the liquid phase and can be burned as such in furnaces or modified internal combustion engines. If water vapour and CO_2 are removed, the remaining methane, about 50 per cent in volume, can, after further purification, be compressed and shipped in natural gas pipelines.

While such process technology might seem to be benign as far as environmental impact is concerned, as with any solid substrate-based microbial process, the question of conversion becomes a major operational problem. High rates of production of methane are incompatible with high levels of substrate conversion, so that slurry handling becomes a critical operating problem. In systems where optimum slurry conversion is a process objective, *e.g.* waste sewage sludge treatment by anaerobic technology, biogas production becomes suboptimal.

Biopolymers

There is a continuing search for advanced materials with novel functionalities. Alternative synthetic routes are sought when these materials cannot be made by conventional means. So far, biological polymers, flocculants and absorbents have been reported, and one bacterial polymer, polyhydroxybutyrate (PHB), has been commercialised. This is a thermoplastic polyester, which should help to alleviate problems associated with the disposal of ''non-biodegradable'' petroleum-based plastics. While the efficacy of this biodegradable product has yet to be fully validated, it would appear to offer a green alternative to some of the persistent organic packaging material presently in use.

Flocculants are widely used for waste treatment. However, synthetic polymeric flocculants, such as polyacrylamide, are not biodegradable and the monomer, acrylamide, is neurotoxic and a carcinogen. There is a clear demand for an environmentally friendly alternative.

Absorbent materials have many uses, for example in baby diapers. A microbially produced bioabsorbent not only absorbs 1 000 times its own weight of water, but, in contrast to synthetic materials, maintains its absorbency in highly saline environments. Analysis of this material reveals it to be a polymer of glucose and glucuronic acid.

Appendix 4

Detection and Monitoring

Traditionally, environmental pollutants have been measured by techniques such as gas chromatography (GC), coupled in some cases with mass spectrometry (MS), high pressure liquid chromatography (HPLC), and ultra-violet spectroscopy. New physical methods which may soon be used include: chemical sensors, x-ray fluorescence spectrometry, and luminescence spectroscopy. This appendix describes the role of biological techniques, specifically the use of biosensors and immunoassays in the measurement of pollutants and the use of genetic techniques to locate and trace micro-organisms in the environment.

Biosensors

The immobilised biological agent in a biosensor may be a nucleic acid, a neural receptor such as a crab antenna, a whole cell, an organelle, a tissue, a membrane, or an entire organism. Most typically it is an enzyme or antibody. The transducer can be potentiometric, amperometric, conductimetric, optical, colorimetric, piezoelectric, or acoustical. Biosensors were produced primarily for clinical and industrial use, but more and more are being developed for environmental applications.

Detection systems, employing biosensor technology, in use or under development, may be classified by type and location of the monitoring process and the distinction between broad spectrum and individual analyte-specific biosensors. On-line monitoring requires the capability of unattended use at source rather than sample collection and delivery to sensor systems in the laboratory. Both continuous "flow- through" and discontinuous sample monitoring may be possible. Off-line systems normally involve sample analysis either in the laboratory or the field, with the latter approach much favoured as it allows far greater flexibility and efficiency of environmental monitoring.

The nature of the biological component determines the specificity and sensitivity of the biosensor. Broad spectrum biosensors exploit living cellular systems or enzyme inhibition by monitoring metabolic or enzymic activity for evidence of perturbation by environmental pollutants; some are capable of detecting a wide range of compounds. For example, algae-based biosensors have been applied as rapid indicators of the presence of herbicides in rivers, lakes, and waste water. The chlorophyll fluorescence emissions of the algae can be heavily influenced by toxic compounds acting as electron transport blockers. Opto-electronic units can immediately detect the emission changes. A range of cellular biocatalysts have been employed, and this type of system offers considerable potential for acute incident detection. Biosensor stability and sensitivity appear to be better with microbial than with enzymic biocatalysts. The use of several biocatalysts, with different sensitivity spectra, may also allow some degree of identification of the pollutant.

Analyte-specific biosensors for environmental use are most commonly targeted at detection rather than continuous monitoring. They are more sensitive than the broad spectrum biosensors and are best used for chronic levels of pollutants. The field assay of biological materials is difficult due to the need for detection of often minute quantities of specific material in the presence of a large background. The most effective analyte-specific systems exploit either enzyme-substrate or antibody-antigen binding, although some use has been made of microbial cells. The development of such systems requires the identification/selection of the appropriate biocatalytic or affinity system for the pollutant in question. Immunological-based sensors are being developed which have the capability to differentiate between analytes and background. Antibodies are bound to either optical or electrochemical transducers, and the substrate-binding event is measured by the generation of an appropriate signal. Typically, fluorescent dyes or chromogenic (colour-producing) enzyme substrates are used for the transduction of an optical signal, while enzyme-substrate combinations which yield electrochemically active species are used with the electrochemical sensor.

Pesticides in contaminated water have now been remotely detected via fibre-optical light guides. One example is based on the capability of organophosphorus pesticides to inhibit acetylcholinesterase; this reaction can be coupled to the formation of a blue dye from a red substrate.

Tests based on whole cells are potentially useful as biological warning systems in the screening for hazardous substances or groups of substances in waterways. Electrodes containing cyanobacteria have been used for immediate detection of acute toxic effects. These electrodes, which have cyanobacteria (*Synecoccus*) affixed to them, allow the measurement of the photosynthetic electron transport system under alternating light-dark conditions in a flow-through cell against reference electrodes. Atrazine and linuron have been detected at concentrations of less than 1µg/l and 100µg/l, respectively.

Current research and development is focused on:

– Broad spectrum devices, capable of on-line use and requiring little or no pretreatment. These can be further sub-divided into systems based on microbial uptake of O_2 or evolution of CO_2, systems monitoring microbial metabolism by mediated amperometry, and systems exploiting enzymic inhibition.
– Analyte-specific devices capable of on-line application requiring little or no pretreatment. Microbial-based biosensors have been developed with good specificity for particular pollutants (*e.g.* phenol). Enzymes using specific pollutants as substrates have also been employed.
– Analyte-specific devices for sample analysis. Biosensors based on the detection of enzyme- substrate interaction or antibody-antigen binding, where the substrate or antigen is the analyte in question, exhibit good specificity and sensitivity. Biosensors of this type have been developed for some carcinogenic hydrocarbons and pesticides.

Immunoassays

Immunoassays consist of specific antibodies and a detectable label on either the antibody or the analyte, with sensitivity determined by antibody affinity and the detection limit of the label. The two basic categories of immunoassays, isotopic (involving the use of radioisotopic labels) and nonisotopic (includes enzyme and fluorescent immunoassays), are based on the type of label used. The antibodies themselves may be polyclonal (complex mixtures produced in animals) or monoclonal (uniform molecules produced by cultures of hybrid cells).

Enzyme immunoassays (EIAs) were introduced in 1971. The enzyme immunoassays most commonly used for analysis of toxic compounds are referred to as competitive enzyme-linked immunosorbent assays (ELISAs). Depending on the assay format chosen, either the specific antibody, the analyte, or a non-specific secondary antibody can be labelled with the enzyme. Although still a novel application of this technology, immunoassays of various types have been developed for environmental pollutants, particularly pesticides. Among these are radioimmunoassays (RIAs) for dieldrin and parathion, and ELISAs for molinate and paraquat. Immunoassays can be easily and inexpensively automated. Many formats are well suited for analysing the large numbers of samples necessary for effective environmental monitoring and surveillance. Immunoassays developed for widely used pesticides are good candidates for automated, high-volume sample analysis. A number of immunoassays which have detection limits as low as 1-5 ppb have been developed for pesticides in water. Some analysts still prefer a combination of GC and HPLC methods where the detection of many pesticides in one procedure is required, and others still doubt whether immunoassays can be an effective alternative, especially since cross-activities have been reported.

Portable field-test immunoassay kits for polychlorinated biphenyls (PCBs) and pentachlorophenol (PCP) are now commercially available. Field test kits have also been developed for the toxic aromatic hydrocarbons benzene, toluene, and xylene.

Immunoassays are usually performed in an aqueous matrix but the water solubility of the target analyte is not necessarily a limiting factor. For example, a rapid, sensitive, enzyme immunoassay for aldicarb was recently developed. Because of its moderate water solubility, instability at high temperatures, and weak absorbance maxima, analysis of this compound by classical techniques is cumbersome.

The toxicity and biological activity of organic compounds is often dependent on their isomeric form. For example, there are eight optical isomers of the insecticide allethrin. Nevertheless, an immunoassay has been developed which is selective for the one isomer which has biological activity.

Analysis of some compounds by immunoassay is not feasible because of the difficulty in raising antibodies to compounds that are highly reactive, very small, very lipophilic or immunosuppressive. Equally, they cannot be used effectively for identification and quantitation of compounds found in complex mixtures or in matrices that contain high levels of interfering substances (*e.g.* metal ions, oil, surfactants).

ELISA assays have been adapted to a variety of analytical procedures in recent years because of the exquisite specificity of monoclonal antibodies. An example is the recognition of immobilised mercuric ions by a specific monoclonal antibody. The use of an ELISA for detection of metal ions compares favourably with the conventional atomic absorption spectroscopy. For instance, samples can be analysed in parallel, enabling large numbers of samples to be processed at one time. In addition, quantitative analysis can be performed with a simple spectrophotometer. Since the assay yields a visible colour change, semi-quantitative procedures can be developed which require no electronic instrumentation for evaluation, giving the assay potential for field use. Finally, the procedure requires only 0.1 ml of sample, up to 1 000-fold less than required by atomic absorption.

The number of monoclonal and polyclonal antibodies raised against pollutants is growing continuously; these will be used in assays but will also be candidates for biosensors once the technical problems associated with their immobilisation on suitable substrates is solved. An example of what has already been achieved in this field is the development of an assay applied to a flow injection system which can be run automatically. Antibodies to atrazine were immobilised on membranes and these membranes were mounted in a cross-flow reactor. Atrazine competed with an atrazine-peroxidase conjugate for the antibody binding sites and the products of the linked reaction were measured fluorometrically.

Gene probe detection of micro-organisms

This technique involves using nucleic acid hybridisation to detect target nucleic acid sequences by binding these sequences with a homologous complementary probe sequence. Most current protocols are based on the hybridisation of a target immobilised onto a membrane. The advantages of this technique is that different types of nucleic acid with varying purity can be analysed and that multiple samples can easily be processed and quantified simultaneously.

The standard method of detecting nucleic acid hybrids is by labelling the probe with radioactive nucleotides and detecting the probe/target hybrids via autoradiography. To avoid possible radiation hazards various non-radioactive colorimetric hybridisation detection methods have been described including the use of chemiluminescent dyes. The non-radioactive colorimetric products have longer shelf-life than the radiochemicals but their sensitivity of detection is lower than radioactive detection methods.

Colony hybridisation

Colony hybridisation is perhaps the simplest application of nucleic acid hybridisation and the easiest to integrate with conventional environmental microbiological sampling and analysis. Bacterial colonies or phage plaques can be transferred from primary environmental cultivation media to hybridisation filters. Growth on the isolation medium increases the number of copies of the target gene to a level that can be detected by a gene probe. This method has been shown to be capable of detecting one colony in approximately 1 million.

The major uses for colony hybridisation in environmental studies have been for the detection, enumeration, and isolation of bacteria with specific genotypes and/or phenotypes. Examples include toluene degraders, naphthalene degraders and mercury-resistant bacteria. Colony hybridisation and gene probes with various specificity have been used to study the maintenance of catabolic and antibiotic resistance plasmids in groundwater aquifer microcosms. Colony hybridisation with a naphthalene gene probe has been used to correlate gene frequency with naphthalene degradation by naphthalene-degrading bacteria in activated sludge. The gene probe analysis for the catabolic genotype is nearly two orders of magnitude more sensitive than the standard plate assay for naphthalene degradation.

In general, for this type of environmental assay the organism of interest must be relatively abundant in the population so that at least one positive colony on an agar plate of 100 to 1 000 colonies can be found. Additional sensitivity can be achieved by plating the isolated bacteria onto selective agar before colony hybridisation. This approach has been used to increase the detection limit for *Listeria* from dairy products.

Isolation of nucleic acids from environmental samples

Two approaches have been used to recover DNA from environmental samples: isolation of microbial cells followed by cell lysis and nucleic acid purification (cell extraction), and direct lysis of microbial cells in the environmental matrix followed by nucleic acid purification (direct extraction).

A simple method for isolating nucleic acids from aquatic samples has been demonstrated which allows filtering relatively large volumes of water. Cell collection and lysis are performed in a single filter cartridge, and chromosomal DNA, plasmid DNA, and RNA can be selectively recovered. One simple but significant improvement of this approach has been the inclusion of a treatment to decrease sample humic content prior to cell lysis, thereby simplifying DNA purification. Direct extraction of DNA from environmental samples was first described in 1987. In this method cells are lysed while still within the soil matrix and the DNA is then extracted with buffer. This method has been successfully used to recover DNA from sediments and soil.

Amplification of DNA – The polymerase chain reaction

The Polymerase Chain reaction (PCR) allows the many-fold replication of lengths of nucleic acid chain, thus amplifying specific segments for subsequent identification. PCR is a useful technique for environmental surveillance of important pathogens. Choice of conditions allows the selective detection of *E. coli, Salmonella* and *Shigella* spp. for example. PCR amplification and radiolabelled gene probes have been used to detect as few as 1 to 5 viable *E. coli* cells in 100ml of water. This demonstrates the potential use of PCR amplification to detect indicators of faecal contamination of water with the necessary specificity and sensitivity for monitoring the bacteriological quality of water in order to ensure the safety of water supplies.

Legionella pneumophila has been detected by DNA amplification and a method specifically for the detection of *Legionella* in environmental water sources has been demonstrated, based upon PCR and gene probes. All species of *Legionella,* including all 15 serogroups of *L. pneumophila* tested, were detected.

Detection of genetically modified micro-organisms (GMOs)

PCR has also been used to amplify a probe-specific region of DNA from an engineered herbicide-degrading bacterium, *Pseudomonas cepacia* AC1100, in order to increase the sensitivity of detecting the organism. Amplified target DNA was detectable in samples initially containing as little as 0.3 picograms of target. The addition of considerable amounts of non-specific DNA isolated from sediment samples did not hinder amplification or detection of the target DNA. The detection of the target DNA was at least a 10^3-fold increase in the sensitivity of detecting gene sequences compared with analysis of non-amplified samples. PCR performed after bacterial DNA was isolated from sediment samples permitted the detection of as few as 100 cells of *P. cepacia* AC1100 per 100 g of sediment sample against a background of 10^{11} diverse non-target organisms; that is, *P. cepacia* AC1100 was positively detected at a concentration of 1 cell per g of sediment.

PCR has permitted the detection of a genetically altered *Es.coli* which was seeded into filter-sterilised lake and sewage water samples. The PCR method amplified and detected the DNA marker of the GMO even after 10 to 14 days of incubation. The PCR method required only picogram amounts of DNA and had an advantage over the plate-count technique, which can detect only culturable micro-organisms. The method may be useful for monitoring GMOs in complex environments, where discrimination between GMOs and indigenous micro-organisms is either difficult or requires time- consuming tests.

Reporter genes and genetic markers

While the variety of genes used as genetic markers has increased with the advent of recombinant DNA techniques and applications, the earliest experiments with genetic fusions made use of spontaneous deletion mutations. One of the main reasons for the success of these fusions is the use of the *lac* operon. It would be difficult to exaggerate the importance of the *lac* operon to molecular genetic analysis, particularly through the use of the *lacZ* gene, which encodes β-galactosidase. In recombinant genetic constructions that are too numerous to mention, *lacZ* fusions have proved their worth as a tool of genetic analysis. As with all good reporter genes, the *lacZ* gene product has an excellent assay. The β-galactosidase enzyme cleaves the disaccharide lactose into glucose and galactose. When the synthetic compound 5-bromo-4-chloro-3-indolyl-β-D-galactoside (X-gal) is substituted, the β-galactosidase is able to cleave the compound, creating an insoluble blue pigment that is very noticeable on most agar media. Interruptions of the *lacZ* gene give white colonies (as opposed to blue) on X-gal-containing media. Successful recombinant fusions appear blue against a background of white non-recombinant colonies. These fusions are particularly valuable as reporters of gene activity when the gene of interest can not be assayed directly. Successful β-galactosidase fusions can also be assayed in liquid medium using o-nitrophenyl-β-D-galactoside (ONPG). This is a colourless substrate which is hydrolysed by β-galactosidase to o-nitrophenol, which is yellow.

The versatility of the *lacZ* system makes it the most widely used reporter gene. It has even been used as a reporter of recombinant *Pseudomonas* in studies to determine whether genetically engineered micro-organisms will survive and disperse if released at an environmental site. *Pseudomonas* strains that incorporate the *lacZ* genes are capable of cleaving the X-gal substrate to produce the characteristic blue colour. This trait makes them very distinguishable from the non-recombinant *Pseudomonas* species that are indigenous to the site.

Other reporter genes are commonly used and deserve mention. All have one or more of the attributes outlined above, and are useful for a specific task. The *gusA* gene, encoding beta-glucuronidase, has been used as a reporter of promoter activity. The gene for chloramphenicol acetyl transferase (CAT) mediates resistance to the antibiotic chloramphenicol. Both a radioactive substrate, ^{14}C-labelled chloramphenicol, and a non-radioactive spectrophotometric substrate have been used in CAT assays. In addition, there are several other fusion vectors available which utilise antibiotic resistance genes, such as for kanamycin and tetracycline. In a similar manner the *xylE* gene of the TOL plasmid has been cloned for use as a transcriptional fusion reporter gene. This gene encodes catechol-2,3-oxygenase, and a simple aerosol spray technique has been used to find recombinant colonies which turn yellow in the presence of the catechol spray. The *xylE* reporter gene is utilised for a quantitative description of promoters from *Streptomyces*, a species for which *lacZ* fusions have not proven effective.

The gene for ice nucleation (*inaZ*) has been isolated and incorporated into a transposon. This transposon was then used to mutagenize plant pathogenic bacterial strains, and the mutants were tested using a droplet freezing technique. Expression of the *inaZ* gene results in rapid freezing of the droplets at –9 C. Dilutions of the cell suspension can be tested in the same way, allowing a sensitivity that is reported to be far greater than the *lacZ* system.

The production of visible light (bioluminescence) by various organisms has been studied in great detail in recent years, and its use as a reporter system has increased with genetic and physiological analysis of the *lux* genes, and with the availability of cloning vectors and other genetic constructions that facilitate their use. Several bacterial species can produce light, including members of *Photobacterium, Alteromonas, Xenorhabdus*, and *Vibrio,* although *Vibrio* has received the most attention.

The first practical application of bacterial bioluminescence was in 1985 when fusions were constructed between *E. coli* promoters and the *lux* genes of *Vibrio fischeri*. Induction of the fusions resulted in a significant increase in bioluminescence, promoting light measurement as a convenient *in vivo* assay for gene expression. Another valuable use of *lux* gene technology was studied in *Streptomyces coelicolor,* where the *lux* genes were fused to a variety of host promoters, some of which were only active during a specific phase of sporulation, or in a specific structure of the differentiating hyphal mass. In another demonstration, the researchers claimed that their construction would produce enough light so that the detection of a single cell's light output was possible. These examples indicate the versatility of the *lux* system, suggesting that its use may be fairly compared to that of the *lac* operon.

Environmental analysis using a bioluminescent reporter strain now appears to be a practical technique. Despite the quenching of bioluminescence that occurs in particulate and contaminated samples, both the presence and bioavailability of specific contaminants were detected in complex slurries. The great sensitivity of this technique has been shown in an analysis of bioluminescent *E. coli* added to soil and liquid media. Results indicate that bioluminescent bacteria can be detected using a non-destructive assay with a sensitivity that is matched only by hybridisation analysis. Multiple reporter genes such as *lac* and *lux* have been transferred into the chromosome of various pesticide-degrading microbes for efficient monitoring after release into the environment.

Appendix 5

Biotreatment of Air and Off-Gases

Filterbed and biofilter design

The principles of biofiltration are based on the aerobic conversion (oxidation) of airborne pollutants by micro-organisms in a water film attached to an organic material carrier. The classical filterbed typically uses peat or soils as filter media and includes appropriate addition of nutrients and buffers (limestone, etc.) to the media to support microbial activity. Water is supplied through nozzles above the media to provide moisture necessary for biodegradation.

Traditional filterbeds have simple designs consisting of:

- an air blower to force the contaminated air through the media;
- an efficient air distribution system;
- a bed of support media (soils, peat compost, etc.) for the microbial films, usually
- operated as a single stage;
- water distribution system (nozzles);
- a water collection system to remove excess moisture.

These filters use inexpensive construction materials but require relatively large areas and low air loading rates. They have been used to control odours and organics in air at low COD loadings (<1 000 mg/m³ ppm) for more than 20 years, generally with successful performance.

Recent biofilter designs involve the use of closed systems. These may be of the form of a biotower with one or more filter sections or stages with individual water distribution systems for each stage. The biotower supports substantially higher air loading, is smaller than the classical filterbed and requires careful operation to prevent process failure. Various proprietary biotower designs are available mostly in Europe.

A second alternative to the traditional design is the BIOVENT system which is larger than a biotower but smaller than a classical biofilter. The BIOVENT system, developed in Germany, has improved air distribution with integrated air supply and drainage ducts and drivable bed support surfaces for mechanised installation and maintenance of the filter media.

The efficiency of gas removal in traditional filterbeds is related to the solubility of the gas components. The surface area available for gas transfer is large and so solubilities can be low. Even non-polar solvents can also be removed by initial adsorption followed by biological breakdown, but flow and loads need to be lower.

Water is also necessary to maintain good microbial activity and evaporation has to be counteracted by water sprays and humidifiers. The water content of the bed needs to be at least 50 per cent. Prehumidification can be with water sprays or steam. If steam is used, the temperature of the reactor has to be carefully controlled to avoid overheating. If the air temperature is higher than 40 °C then cooling is necessary either by dilution with fresh air or by the humidifiers. There will be some heating between 2-4 °C because of the exothermic nature of the oxidation reactions.

Modern filter beds are moving away from natural materials such as peat towards synthetics with high surface area, typically polyurethane foams, porous clay and polystyrene beads to optimise durability, reactive surface and reduce back pressure. A proprietary filter with a bed life of five years has permitted loadings of 300-500 m³/hour as compared with the 30-50 m³/hour for a traditional filter.

With synthetic materials it is possible to construct multi-staged filters each with mixed cultures of micro-organisms selected to degrade specific groups of pollutants.

Microbiology

Various studies have been made of biofilter microbiology. The most active organisms have been found to be heterotrophic and chemo-organotrophic, with *Actinomycetes* species one of the most common groups, able to degrade a wide variety of organic compounds. Bacteria and fungi are also active. Organisms are found to be vertically stratified with the highest population densities in the lower portions of the bed. It is assumed that the most easily degradable organics are consumed first and the more refractory compounds metabolised in the upper portion of the bed. Significant reduction of the organic concentration of contaminated air is achieved within approximately the first week following start-up with maximum reduction in the first month. Acclimatised biofilters can be operated with intermittent air flow without significant loss of performance. A summary of various studies suggests that the following conditions are optimal:

- temperature   15-45 °C
- bed   pH 7-8
- bed moisture content 50-70 per cent
- influent air humidity 80-100 per cent.

Micro-organisms have a tremendous ability to adapt to new substrates. Once a new strain or mixed culture has been isolated in the laboratory, it may generally be applied directly to the filterbed. Easily biodegradable compounds can be degraded by the organisms found in bed material but xenobiotic compounds may require the bed to be inoculated with specially cultivated organisms. Examples of these are species of *Nocardia* for degrading styrene, *Hyphomicrobium* for dichloromethane and *Xanthobacter* for 1,2-dichloroethane. Nutrients may have to be added to the bed in addition; for example, toluene degradation has been shown to improve significantly in this way.

Performance

Levels of pollutant removal found in a number of studies are shown in Table 6:

A limitation on existing technology is that the biofilter operation is an unsteady state process. In it, the residual biomass from biodegradation accumulates in the filter media. Degradation of chlorinated organics and sulphur compounds produces acids (HCl and H_2SO_4) that consume the buffering capacity of the bed. Thus the process, at high organic loadings in air, experiences increasing pressure loss across the media and decreasing process pH. At all pollutant loadings, the process eventually requires replacement of the support media. The low pH that can occur within the process can damage the system's construction materials if these are not acid-resistant. With unsteady state operation, the time of operation before media replacement depends upon the pollutant loadings. Low pollutant loadings have resulted in process runs of as long as five years although one to two years is more usual. In the US EPA's Risk Reduction Engineering Laboratory in Cincinnati, Ohio, high pollutant loadings (>200 ppm) of organic compounds in air were fed to a biotower. The biotower removed essentially all of the organic pollutants from the air in two minutes of retention time but exhibited relatively rapid accumulation of residual biomass and tower flooding in several months. The chlorinated organics also produced stoichiometric demands for buffer.

The concentration of organics in contaminated air at industrial plants or stripped from contaminated leachates or groundwater can reach several hundred ppm of organic compounds in air. Under these loadings, a

Table 6.   **Efficiency of removal of pollutants from air by biofiltration**

Substance	Per cent removal
Hydrogen sulphide	99
Dimethyl sulphide	91
Terpene	98
Organo-sulphur gases	95
Ethyl benzene	92
Tetrachloroethylene	86
Chlorobenzene	69

Source: R. Atlas, University of Louisville, United States.

steady state operation of the biofilter, with removal of excess biomass and continuous mineralisation of the produced acids, is highly desirable. The US EPA's Risk Reduction Engineering Laboratory is now developing steady state, low pressure drop (> 6 cm of water column) systems with effective buffering capabilities. The improved design concepts are currently being evaluated for patentability.

Appendix 6

Land and Soil Bioremediation

Technologies for bioremediation fall into two basic categories *in situ* treatments and *ex situ* treatments.

In situ treatments

These methods are designed to encourage pollutant degradation by micro-organisms present in the soil without requiring soil removal (Case Studies 15 and 16). Nutrients are added by irrigation or are pumped into the polluted area. Introduction of air or pure oxygen is difficult and a number of strategies have been adopted.

   a) Injection – lances are used to inject air from a compressor into the contaminated ground at various pressures. This is an effective technique, though it can be quite operator intensive as each lance only has a limited sphere of influence depending on the nature of the soil, and a large number of lances are needed to treat a relatively small area.

   b) Buried Pipes – pipes are laid horizontally below and within the contamination plume and air is passed through from a compressor or blower. Again this is an effective technique and though the capital cost is higher than for the lance system it has lower operating costs.

   c) Aerated Water Circulation – water is abstracted and aerated vigorously in tanks and reinjected, the theory being that the water can carry oxygen through the contamination plume. This technique is limited by the solubility of oxygen in water which is very low. Attempts have been made to increase the oxygen loading in water by adding hydrogen peroxide which doubles the effective concentration. This is a useful adjunct to the aeration methods because it enables additions of nutrients and microbial inocula to be made, but it is not enough in isolation.

   d) Recirculation wells – pumps at predetermined depth inside the well casing circulate water from one level to another. Air or peroxide solution may be added via the pump.

Ex situ treatments

Technologies for *ex situ* remediation include landfarming and the treatment of soil in specialised bioreactors.

Landfarming

Waste residues are applied to the surface layers of a soil and the soil is cultivated regularly to ensure adequate mixing and aeration. Aerobic microbial activity in the soil then ensures the degradation of the waste. Soil is laid out to a depth that can be cultivated using conventional agricultural equipment (usually < 0.75m), and chemical and microbiological treatments are applied at intervals during the treatment period. The soil bed is usually constructed above an impermeable base, incorporating a leachate collection system.

The application of contaminant-degrading micro-organisms to the soil may be required to establish an appropriate population of micro-organisms. The latter may involve the creation of an aerobic microbial community from what was originally predominantly anaerobic when the soil was buried at depth. In some instances it is also beneficial to add inert materials to the soil treatment bed in order to enhance the soil structure and improve aeration.

Single treatment regimes often result in an exponential degradation, as the more biologically degradable components are degraded first. However, there is usually the option to enhance the degradation of the more

Case Study 15:   In situ remediation of contaminated land and groundwater – United Kingdom

A contract to undertake the remedial treatment of an oil spillage which had contaminated soils and groundwater involved the application of *in situ* treatment methods to reduce oil concentrations to safe levels, avoiding the need for the mass excavation and disposal of materials.

On the site, a significant quantity of heating fuel had leaked from a storage tank over a period of a few days due to damage to a pipeline. The fuel had migrated down through relatively porous ground and polluted a groundwater aquifer. The oil had dispersed on the surface of the water table resulting in a spreading plume across an area of over 15 000 sqm. Furthermore, natural vertical movement of the water table resulted in the oil becoming smeared within the vadose zone.

A system was installed to control any further migration of the oil plume and to treat contaminated soils and groundwater to acceptable concentrations.

The mass excavation and disposal of soils was ruled out owing to the large volumes involved and potential difficulties of excavation and dewatering around buildings and services. An *in situ* approach using enhanced bioremediation and vacuum extraction was therefore adopted. In addition, an oil/water separator was installed to prevent any further contamination of nearby surface waters.

The *in situ* treatment system consisted primarily of underground feeder/extraction pipes used to control a network of well points. Three categories of boreholes were installed, including groundwater monitoring wells. *In situ* remediation of the main oil plume involved containment and treatment using a combination of enhanced bioremediation of soil and groundwater, and vacuum extraction within the vadose zone. Stimulation of the indigenous oil-degrading soil microflora was achieved by forced aeration of the saturated and vadose zones.

Upwelling was alleviated by controlling the rate of air injection and by utilising pulsed vacuum with intermittent forced air injection. This was important in order not only to contain groundwater flow but also to prevent contamination of the near surface strata.

The site was continually monitored over the six-month treatment period. Final validation confirmed that oil concentrations in soil had been reduced by over 98 per cent and the concentration of potentially toxic organic components including benzene, toluene and xylene were undetectable in both soil and groundwater.

Source: Celtic Technologies Ltd, Cardiff, United Kingdom.

Case Study 16:   Oil contamination in the Upper Rhine Valley – Germany

A mineral-oil spill occurred in an industrial area of a city in the Upper Rhine Valley. Although free-phase hydrocarbons were removed from the subsurface by pumping over a period of three years, residual saturation of the hydrocarbons remained in the groundwater. The high residual concentrations of aromatic hydrocarbons formed a plume which was moving towards a municipal waterworks. Remediation was needed to inhibit plume migration and to clean up the groundwater.

Several sandy aquifers present in the area are separated from each other by extensive impermeable layers. The upper aquifer thickness varies from 8 to 10m and primarily consists of sand with very little silt and clay. The clay layer below the upper aquifer is 10 to 30m thick and prevents downward migration of contaminants. The municipal waterworks are located at a distance of 500m from the spill location along the direction of groundwater flow.

Groundwater withdrawn from the deeper (clean) aquifers was used to create hydraulic barriers for preventing the spread of the plume to uncontaminated areas. Nitrate was added to flushing water as an electron acceptor and phosphate and ammonia were added to optimise the bacterial growth conditions. Since substantial depletion of nitrate at the midplume locations was observed, the flushing water was heated from the ambient 12 to 22 °C prior to injection. Heating the flushing water helped increase both the growth rate of the micro-organisms and the solubility of the hydrocarbons.

The total microbial cell count was 2×10^7 per ml of flushing water. Before the restoration operation, the cell counts were three orders of magnitude lower. Groundwater present outside the plume showed similar low cell counts during the restoration operation. High carbon dioxide and molecular nitrogen content in the flushing water indicated hydrocarbon mineralisation and denitrification.

The concentrations of aliphatic and aromatic hydrocarbons declined from initial levels of 2.0 and 6.0 mg/l after an operating period of six months and thereafter steadily declined to 0.1 and 0.08 mg/l in two years. From the aromatic fraction, all the benzene was mineralized in six months, with toluene degradation occurring thereafter. Xylene degradation was slow in the beginning; however, it steadily improved once benzene and toluene disappeared from the system.

Source: Environmental Protection Agency, United States.

Figure 15. **Diagram of a landfarming operation**

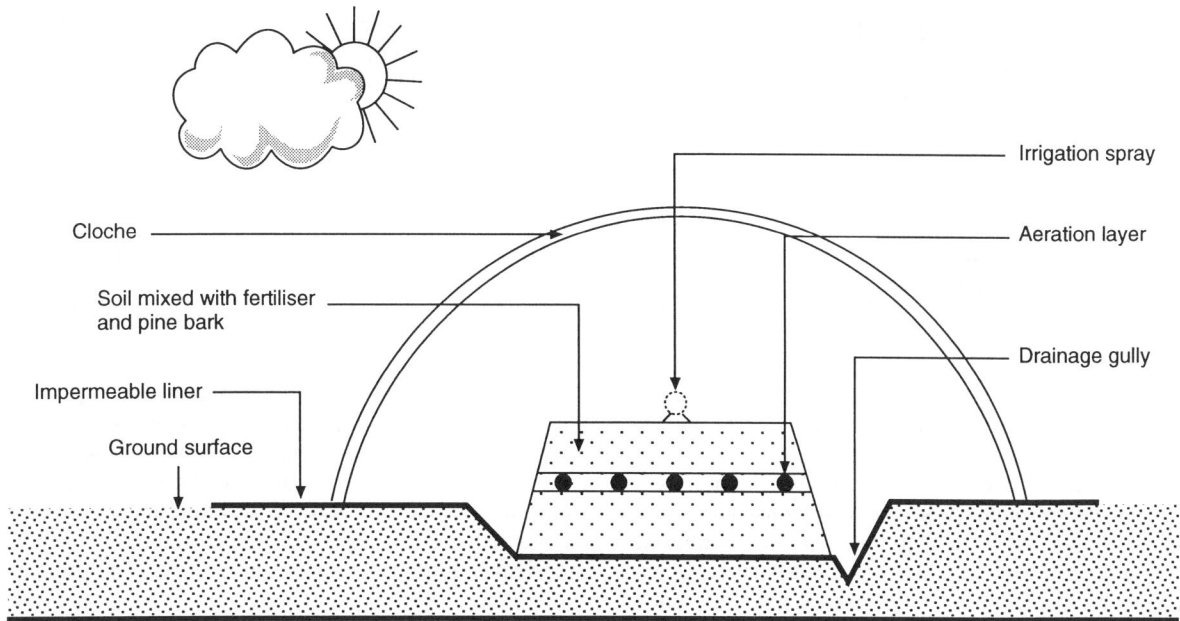

Source: Shell Research Limited, United Kingdom.

recalcitrant residues at a later stage by optimising the treatment for those compounds. Such optimisation may include the application of surfactants, co-metabolites or microbial degraders specific to the remaining residues.

Contaminated soil may be moved to a selected landfarming site. Here it can be covered over with a plastic greenhouse, water circulated to increase the temperature, and nutrients added (Figure 15). In this fashion, the process is largely independent of the weather and should take no more than three to six months to reduce, for example, oil levels from 5 000ppm to 100ppm. There is no need to augment the naturally occurring organisms in this system although it is often done anyway as insurance.

Soil bioreactors

Ex situ treatments may involve the use of slurry bioreactors. Sizes of bioreactors vary from small skid-mounted units with a working volume of 1-5m^3 to large custom-built plant for the treatment of large sites. Unless dedicated plant is built, there is a heavy constraint on the treatable volume in spite of the short retention times. Materials to be introduced into a slurry reactor may be pre treated in a hydrocyclone to separate the pollutant-bearing fines from coarser materials.

The operating procedures most commonly employed reduce gross macroscopic heterogeneities by mixing and by the introduction of water to form aqueous slurries, thereby replacing air with water as the continuous phase within the bioreactor. Even so, numerous transients and gradients will continue to predominate. A broad range of physico-chemical factors have important impact on the operation of bioreactors, including the hydrodynamic properties of suspended microbes; electrokinetic properties of microbes, surface, wall and interface effects; multi-phase flow effects; effects in foam, froth and spray environments; flotation, sedimentation and segregation effects.

Problems faced in the bioengineering of effective and efficient bioremediation processes include the scarcity of data concerning the physiology of the micro-organisms, including bacteria, archaebacteria, actinomycetes, yeasts and filamentous fungi, that have potential for employment in either *in situ* or bioreactor remediation operations. There is also an inability to define the highly heterogeneous, continuously varying process conditions that are encountered in either type of operating systems. The ultimate performance criterion in any bioremediation process is usually the residual pollutant levels that can be achieved, rather that the percentage of the pollutants that have been removed.

Solid-phase treatments may also be used where the fundamental objective is to apply agitation or mixing in order to enhance process rates and to reduce any gross segregation within the bioreactor charge. The flow characteristics of particulate and granular matter in mixed solid phase reaction systems is complex. Little is known of the dynamics but the basic facts show that a granular mass is neither a fluid nor a solid as ordinarily understood. For example, it is a property of a granular mass, in a vessel from which it is being discharged from the bottom, that unless the vessel is nearly empty, the flow rate of material from the vessel is independent of the head of material in the vessel.

A major problem in moving granular beds where microbial growth is occurring is the physical effects of the movement of the granules on the growing microbes. The microbes that are of potential value for bioremediation are not restricted to bacteria, but also include both filamentous fungi and actinomycetes, and these need to be protected from all forms of stress that are detrimental to process operation.

Some stresses that occur during transient state operation can result in either improvement or loss of process performance. Most bacteria have evolved mechanisms that allow rapid growth under favourable conditions, while also permitting survival under conditions that are unfavourable for growth. The response of a growing bacterium to perturbations due to an external stress is first, to cease the current mode of growth and metabolism; second, to protect the cell against any adverse consequences due to the initial response; third, to develop and implement strategies that can deal with the changing conditions. In such a sequence the specific effects involve both gene expression and enzyme activity. Clearly, the growth and survival requirements can be in conflict. The proteins necessary for balanced growth are usually present in requisite quantities, whilst those required for protection, repair or growth under adverse conditions are usually only present in very low concentrations and are synthesised when needed.

The most extensively studied response to a potentially adverse environmental change is the seemingly universal heat shock response which involves rapid transient synthesis of a specific set of stress proteins. However, various other changes or agents also invoke a stress response. Of particular relevance in the context of bioremediation are stress protein induction by substrate and nutrient starvation, anaerobiosis, hydrogen peroxide challenge, osmotic (salt) effects, ultraviolet light, cold, acid (pH) and heavy metal salts.

Site-related issues

Site characterisation

Site characterisation is critical for selecting the most appropriate remedial technology, and for successful implementation of *in situ* bioremediation (Case Study 17). To improve the state-of-practice in subsurface bioremediation, field demonstrations have to apply geophysical tools to define the three dimensional association of the contaminant and associated geological materials. For example, the hydraulic and pneumatic conductivity directly control the flows of nutrients and oxygen into the contaminated zone. Variability in conductivity can leave behind pockets of material that resist remediation while the rest of a site is cleaned up. The capacity of geological materials to sequester and limit the passage of nutrients such as ammonia or phosphate often complicates the implementation of bioremediation.

Typical soil micro-environments include three principal inorganic particulates: sand, silt and clay; a chemically and physically diverse organic fraction, with both natural soil components and a frequently diverse range of pollutants; an aqueous phase, soil water and groundwater, that is frequently discontinuous and contains unevenly distributed, dissolved and dispersed organic and inorganic matter; a gaseous phase of markedly different composition from the overlying atmosphere; and a diversity of both indigenous and non-indigenous micro-organisms (see Figure 3). The structure of soils, and the underlying cohesion of the soil particles into a defined matrix, is controlled to a greater or lesser degree by soil organic matter, and particularly by organic materials, such as extracellular polysaccharides, that are produced by soil micro-organisms.

Case Study 17: Physical and biological treatment of deep diesel-contaminated soils – United States

A ruptured pipe at a Burlington Northern Railroad fueling pump house resulted in over 60 000 gallons of diesel fuel spilling onto the surrounding soil. An initial investigation of site conditions indicated that subsurface soils were contaminated with diesel fuel to groundwater, which was observed approximately 70 feet below ground surface. State regulatory agencies requested that the railroad develop and implement a remedial action plan to treat these diesel-contaminated soils and protect local groundwaters. A company retained for this work recommended using soil venting methods to enhance the immediate volatilisation and long-term biodegradation of fuel residuals, and designed and implemented a pilot test to determine the feasibility of this approach. The objectives of this test were to determine soil properties such as air permeability and to assess the potential for partial volatilisation and long-term biodegradation of diesel residues at the site. Hydrocarbon concentrations, carbon dioxide, and oxygen levels were monitored at a vapour extraction well and six vapour monitoring points. These data were evaluated to determine the rate of volatilisation and biological degradation of fuel residues. A more focused and controlled remediation of deep soils was also an objective of this test. A discrete-level vapour extraction device was used to increase soil gas flow and deliver oxygen to deep contaminated soils.

The diesel fuel contamination in the soil appeared to be localised within a 60 ft radius of the pump house. Based on soils analysis and observations made during drilling, it appeared that diesel fuel migrated rapidly downward at the spill site until it encoutered the interbedded sand and silt/clay zone at approximately 30 to 35 ft. At this depth, the fuel began to spread laterally and continued its downward movement through more permeable sand lenses.

The projected capital and operating cost for remediation of the 25 000 cubic yards of contaminated soil at this site is $5 to $7 per cubic yard for the first two years and an additional $0.25 per cubic yard for each additional year of operation. Economy of scale has a significant impact on unit costs. For example, a diesel spill contaminating only 5 000 cubic yards of soil would require essentially the same capital investment and have only slightly lower operating costs. Even at $25 to $35 per cubic yard, this technology would offer a very simple and cost-competitive solution to many fuel contamination problems.

Source: Engineering-Science Inc., United States.

State of contaminants

The prospects for successful bioremediation are sensitive to the physical/chemical state of the contaminant. Contaminants in solution in water, as well as vapours in air, are relatively easy to treat as are contaminants that sorb to organic matter or mineral surfaces, but desorb readily into water or air. Contaminants that desorb slowly are more difficult: bioremediation can remove the contaminant from groundwater or soil air, only to have the contamination appear to come back after bioremediation ceases. Contaminants that appear as a discrete oily phase, as ''free product'' or tars, are the most challenging. The contaminant is physically separated from the micro-organisms and the rate of remediation is the rate of mass transfer from the source of contaminant to the micro-organisms. Much current research in bioremediation is exploring the use of surfactants to overcome problems with physical state. Surfactants will be used to facilitate desorption, and to break up oils and tars to increase their specific surface area.

Organic solvent wastes are distinguished by their relatively high volatility in air. This property of these wastes discourages remediation methods in above-ground bioreactors because moving the soils would produce large volatile emissions and *in situ* methods are preferred.

Concentration of contaminants

Effective bioremediation can be limited by too little contaminant, or too much. For example, the water solubility of many polynuclear aromatic compounds is too low to provide the sole nutrition for degrading organisms and these compounds only degrade when other substrates are present. The same is true of the reductive dechlorination of PCBs. Research to overcome the problem of low solubility and to promote degradation concentrates on the use of surfactants.

Bioreactors tend to be less effective at the low concentrations of contaminants often seen in groundwater or leachates, or wash water from soil washing, or vacuum extraction systems. Pretreatment techniques that concentrate contaminants prior to ultimate treatment in a bioreactor include soil treatment to segregate fines followed by concentration of the fines in a hydrocyclone, the use of membranes to concentrate solutions of organic compounds or the use of molecular sieves to trap and concentrate vapours prior to desorption into a biofilter.

Water availability

Of great importance as far as bioremediation is concerned is the question of water availability. All microbes require water for growth, but in spite of this, microbial growth can occur in environments that exhibit a wide range of water availabilities. Water availability does not necessarily depend on the total water content in any particular environment, but is a complex function of both sorption and solution factors.

Water availability is a major selective factor with respect to microbes in soils. For movement in a solid matrix, microbes generally require a continuum of water-filled pores of requisite diameter to provide appropriate pathways although most filamentous microbes have the ability to bridge air-filled pores and to penetrate semi-solid matter. This is clearly evidenced by greater domination of soil environments by the latter microbes as water content decreases. This has important implications as far as *in situ* bioremediation is concerned because of the requirement to enhance the biodegradation capacity of the indigenous microbial population during operation and the potential of filamentous microbes in *in situ* bioremediation processes.

Further, pollutant migration during *in situ* operations and partial biodegradation of pollutants and/or conversion to form either hazardous or toxic intermediates, possibly of a different physical state must be considered. In the latter context, particular problems can be anticipated when either gaseous products or volatile liquid products are formed.

Pollutant identity

Prior to evaluating the potential for the effective bioremediation of any particular polluted site, it is first necessary to identify not only the predominant pollutants that require removal from the site, but also to identify noxious minor pollutants that might remain as either modified or unmodified residues at the conclusion of the clean-up operation, thereby rendering the site inadequately restored in law.

Essentially, three major categories of polluted subsurface environments exist. Those that have been polluted by either spillage or leakage of complex mixtures such as crude oil, heavy fuel oil, kerosene, coal tars, etc.; those polluted by a restricted number of defined chemicals, such as solvents or explosives, during their production, subsequent handling, or use; and those that have been used as disposal sites for diverse wastes, including unwanted (banned) noxious chemicals, off-specification chemical products, and chemicals that have been subjected to ill-defined modifications for example, by fire damage.

In addition, it is also necessary to consider a number of ancillary technical requirements that must be satisfied for process operation to be acceptable. The first is that the pollutants present must ultimately be biodegraded to non-toxic stable end products; the second is that neither substrates nor nutrients that have been added to stimulate biodegradation must remain as either residues or potentially unstable intermediates at the conclusion of the clean-up operation; and the third is that the operating practices adopted must not result in the migration of either the original pollutants or any intermediates and/or products formed into other environmental compartments. Regarding the first and second requirements, residual microbial biomass is treated as an exception if it is free from sorbed pollutants and does not contain potentially transferable genetic sequences that have resulted from genetic manipulation, but in the context of the third requirement, it is also important that no significant transfer of microbial biomass between environmental compartments, particularly from soil to groundwater, occurs. Overall satisfaction of these several requirements limits freedom in process operation.

Appendix 7

Water Treatment Systems

Waste water first requires some preliminary mechanical and physical treatment to remove coarse solids and immiscible liquids. After extracting gross particulate matter (cloth, wood, residual metal, etc.) by coarse screening, grit is removed by settlement. The sewage is then passed to sedimentation tanks for the separation of most of the suspended solids as a sludge, thereby eliminating much of the biological oxygen demand. Many effluents also contain significant quantities of large organic solids; although biological reactors are designed to cope with and treat solids it is usually cheaper to settle these out and treat them separately. Settlement of vegetable, meat and domestic wastes can remove 35 per cent of the total pollution load by precipitation of 60 per cent of the suspended solids. Wastes containing predominantly colloidal solids, *e.g.* drinks and dairy effluents, are unlikely to benefit from settlement.

This appendix describes in further detail some of the subsequent biological processes and systems available for the clean-up of both municipal and industrial waste water. Processes are differentiated by fixed film and suspended systems.

Fixed film systems

The three main reactor configurations for fixed film systems are downflow, submerged filters, and fluidised beds.

Most aerobic systems for treating municipal waste operate in a **downflow** mode. The water/wastewater is distributed over the top of the medium supporting the biofilm and flows down due to the action of gravity. In most aerobic systems, the hydraulic loading needs to be relatively low to allow for the passage of air up through the filter.

The earliest type of reactor was the trickling filter, which developed directly from experiments carried out on the treatment of sewage by percolation through soil. They are circular or rectangular tanks made from brick, concrete, or coated steel. They are easy to maintain, incur low running costs, and have a plant life of between 30 and 50 years. Percolating filters have proved to be very reliable, and 70 per cent of European treatment plants use this type of system. In trickling filters the waste water is sprayed over the surface of the filter by nozzle or moving distributor and allowed to percolate down through the bed. Good distribution of the waste over the surface of the filter is very important to provide intimate contact with the biofilm covering the support. The most popular system is a circular filter and a rotating distributor. Usually the force of the water issuing from the holes is sufficient to drive the distributor around and no further power is required.

Traditional aerobic trickling filters use cheap support material such as blast furnace slag and coarse gravel. This type of medium has a relatively low surface area to volume ratio and a low voidage, the hydraulic retention time within the medium is good, and it is cheap and resistant to extremes in weather. Modern plastic media, either randomly packed or designed with a tortuous liquid flow path, have higher surface area to volume ratios and greater voidages. Greater amounts of biomass per unit volume can be supported and still provide enough air for treatment. However, the hydraulic retention time is lower so that filter beds have to be much deeper (15-20m compared to 2m) for effective treatment. Plastic media are more suitable for high-rate treatment of high-strength organic waste waters (BOD >1 000mg/l) from industry, as opposed to weaker municipal sewage (BOD ca.200mg/l). There is a continuous cycle of biofilm growth and detachment taking place, partly due to autolysis and partly due to the action of the grazing organisms. The debris or humus flushed from the biological filter is collected in a final settling or humus tank.

The two characteristics of media which most affect the efficiency of biological treatment are the specific surface area available for biological growth and the voidage for ventilation and drainage. Plastic materials can be moulded into shapes with optimum surface area and voidage, but to be effective, the much greater surface area needs to be uniformly wetted. Two types of plastic media are used, modular self supporting blocks or random rashig rings. Tall biotowers with lightweight prefabricated walls (usually 6m high) are possible with the modular plastic media. The space or voidage between the support medium controls the ventilation and therefore the oxygen available for microbial growth, drainage, and discharge of solids. Ventilation through the medium is assisted by the difference in temperature between ambient and the waste water and by air shafts around the circumference of the filter.

Treatment efficiency is, in principle, proportional to the amount of biomass, but excessive growth of biofilm in biological filters can fill the spaces in the packing material. The amounts of waste that can be treated and the aeration or ventilation required to ensure reliable treatment were at first established empirically. The surface of the filter receives and removes the greatest amount of the applied substrate. The most active organisms are in the top 0.5m of the bed (known as the heterotrophic layer); below this is the autotrophic zone where the bios is mostly debris and humus from the upper more active layers. There is little BOD removal in these layers, but considerable autotrophic nitrification. Higher rate wetting via recirculation increases efficiency by spreading the load deeper into the bed. Dilution of the waste by recirculation improves mixing, reaerates the waste, and recycles valuable nutrients.

Partially or totally submerging the biological support media significantly improves the efficiency of biological filtration. In **submerged** or **flooded** aerobic filters, effluent flows down through a bed of small natural or plastic media and air is injected into the base of the bed (Figure 16). This allows the wastewater to be applied at a much higher rate than conventional filters. There may be a second filtration stage below the injection point to remove solids generated in the upper aerobic mixed zone. The solids, which accumulate in the lower filter stage, eventually reach a critical concentration, and the filter is then cleaned by back-washing and air scouring. Repeated backwashings are used until the system is clear and free of solids. The wash water is returned to the front of the treatment plant for settlement.

In other submerged systems, rotating discs and drums may be used as a base on which the biofilm grows. The weight of the rotor is supported and the rotation driven by the flow of air from the centre to the outside.

In **fluidised bed reactors** the support materials – pumice, sand or plastic – are fluidised by pumping waste water up through the medium (anaerobic) or by injecting air or oxygen (aerobic). In the first instance, the upflow velocity is high enough to fluidise the support particles so that they are not in direct contact with each other but not so high as to shear off all the biofilm. The advantage of this system is that, because the particles are not in direct contact with each other, there is a greater surface area available for biomass support. However, the energy costs for providing a high enough flow rate, usually involving a high recycle rate to fluidise the bed, are greater. Fluidisation can range from 10 to 100 per cent of the bed volume when stationary. The term ''expanded bed'' is sometimes used for fluidised beds with a low fluidisation.

Other suspended growth systems include ''upflow sludge blanket'' reactors, which use ''granules'' of anaerobic bacteria on a solid matrix in an upflow reactor. The settling velocities of these granules is high enough for them to stay in the reactor, with the help of baffles, while allowing the effluent to exit at the top of the vessel. High strength organic waste waters (BOD>1 000mg/l), in particular from food manufacturing, are amenable to treatment in these systems. A further development of this reactor is an up-flow fluidised bed reactor where the granular floc is fluidised (without supporting particles). This permits the building of a slender tower reactor with a small ''footprint'' (Figure 17). The key to the successful operation of these reactors is that the biomass has a retention time much longer than that of the liquid.

Fluidised beds combine the advantages of the percolating filter process with those of activated sludge. Biological fluidised beds are able to retain about five times the concentration of micro-organisms compared to conventional plant and significant savings in capital cost are likely.

Suspended growth systems

The most well-known suspended growth system is the aerobic treatment referred to as the activated sludge process. In its simplest configuration, the activated sludge process comprises two clearly identifiable stages, the aeration tank(s), where the bio-oxidation of biodegradable pollutants occurs, and the clarifier, where the flocculent process microbes and other solids sediment to form a concentrated sludge, part of which is recycled to the aeration tank to enhance intensity and maintain capacity and part of which is wasted. The process is operated in the continuous flow mode with the flocs of process microbes distributed throughout the water undergoing treatment. When activated sludge processes are used for the biotreatment of industrial waste waters, the overall process objective is to achieve treated water discharge standards that either meet those prescribed for direct

Figure 16. **Simplified diagram and photograph of a modern biological aerated flooded filter (BAF) waste water treatment plant**

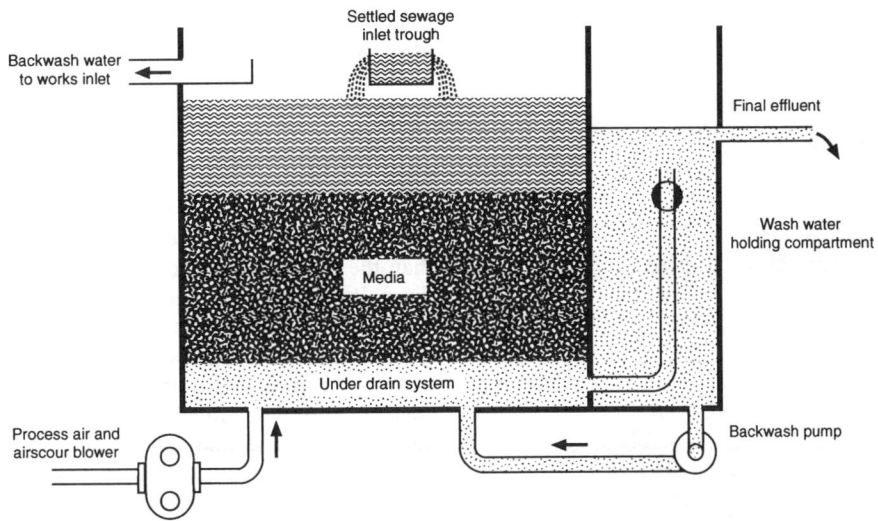

Source: Thames Water, United Kingdom.

Figure 17. **Biobed upflow reactor**

Source: Gb Biothane International, the Netherlands.

discharge into natural surface waters or are such as to allow subsequent tertiary treatment processes to be effective. Detailed process objectives include biodegradable carbon removal and, where appropriate, nitrogen removal, while the process discharge should be essentially free from suspended solids. By-product sludge production should also be minimised. The retention time in the aeration tank is between six and fifteen hours, after which the effluent is separated from the active sludge by settlement (two hours) and most of the sludge is returned to mix with the incoming effluent.

Process microbes are present as flocs which comprise a complex matrix of not only the microbes, but also extracellular, microbially produced, biopolymers. Generally it is considered that it is biopolymer, rather than the microbial surface, that is in direct contact with the polluted waters undergoing treatment. The composition of the biopolymers comprising the non-microbial fraction of flocs depends on both environmental factors and on substrate and nutrient availability. The flocs are suspended in the aerated liquid rather than attached to a solid support. The sludge increases in bulk as the incoming organic material is degraded; some is drawn off to be degraded anaerobically while the remainder is recycled through the system.

Air is supplied to the activated sludge process via submerged diffused aerators, or via surface aerators that produce liquid droplets that are thrown through the atmosphere. The oxidation or Pasaveer ditch is a form of extended aeration in which the sludge is surface aerated in a "racetrack" configuration reactor.

In an activated sludge, chemo-organotrophic and mixotrophic bacteria function as specific pollutant degraders. Remarkably few definitive studies have been made of the predominant genera and species that comprise the biodegradative capacity. Microbiological investigations of activated sludge processes emphasise the roles of biopolymer-producing microbes in floc formation, of filamentous microbes in bulking, scum and foam formation and of protozoa in maintaining low suspended solids contents in the treated water discharge.

Probably significantly more than 80 per cent of the species present in the microbial community still evade isolation and identification. In view of both the broad range of pollutants that can be biodegraded and the continuously changing environmental conditions imposed during biotreatment, the process culture can be assumed to comprise a dynamic mosaic of interacting consortia. In spite of the fact that activated sludge processes were first introduced for municipal sewage treatment more than 70 years ago, and the fact that they are the largest and most widespread microbial processes in existence, little has been done to fully optimise their microbiology.

The anaerobic system analogous to the activated sludge process is called the "contact" process. Since anaerobic consortia of micro-organisms do not readily form flocculent particles sedimentation aids such as polyelectrolytes need to added. The membrane anaerobic reactor is a suspended growth reactor in which a highly active biomass concentration is achieved through the use of ultrafiltration for solid-liquid separation and subsequent biomass recycle to the reactor. The use of ultrafiltration prevents any loss of suspended growth into the effluent and permits the control of the sludge retention time. The reactor is favoured in general for the treatment of low volume, high strength waste waters, containing both soluble and particulate organics.

Developments to the aerobic suspended growth reactor

One of the earliest modifications, to reduce aeration costs, was tapered aeration. Most activated sludge plants are plug flow and the oxygen demand at the outlet is much less than the inlet, so aeration can be progressively reduced along the length of the tank. This is accomplished by using fewer diffuser domes in the base of the tank. Most large activated sludge plants now include an anoxic denitrification stage. There has also been research which shows that an anaerobic stage can be used to increase phosphate uptake by the micro-organisms.

Oxygen availability in the activated sludge process can be increased by using pure oxygen instead of air (Figure 18). There are two proprietary systems: one uses completely sealed tanks and is very similar to conventional activated sludge; the other uses an open tank with external recirculation pumps to keep the tank mixed. High-pressure oxygen is injected into a venturi throat on the outlet side of the pump. This generates a stream of very small bubbles of oxygen.

An alternative method of increasing oxygen solubility is to increase the hydrostatic pressure in the reactor by building tall reactor tanks (Case Study 18). One variety of these, the ICI Deep Shaft, is a tube sunk between 50-100m into the ground (Figure 19). Dissolved oxygen and turbulence in the tube are high and permit a high concentration of very active biomass to be obtained. To avoid flotation the treated effluent has to be vacuum

Figure 18. **Union Carbide's "Unox" process for waste water treatment uses oxygen instead of air for higher oxygen transfer rates**

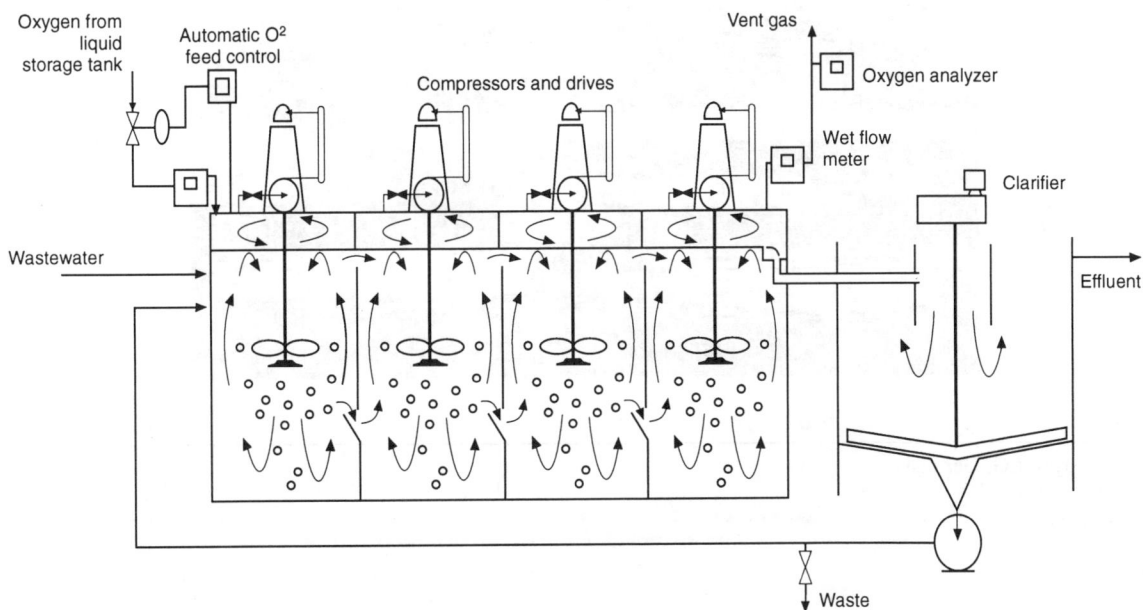

Biological reactor		Secondary clarifier	
Liquid depth	5 ft 2 in (1.6 m)	Diameter	7 ft (2.15 m)
Stage volume	400 gal (1 800 l)	Side water depth	9 ft (2.75 m)
Total liquid volume	1 600 gal (7 200 l)	Center well diameter	26 in (0.66 m)
		Overflow area	33.2 ft² (3 m²)

Source: R. Atlas, University of Louisville, United States.

Case Study 18: Biological Sewage Purification by means of the Bayer Leverkusen-Bürrig tower biology system

The Bayer tower biology system constitutes the most important part of a joint sewage treatment plant in the City of Leverkusen.

In sewage disposal plants using biological techniques, the time required to complete the process is considerably reduced by introducing large amounts of bacterial sludge into the sewage and then adding oxygen to this mixture artificially. The Bayer system does not use conventional, large, open clarifiers but instead closed tanks of up to 30m in height. The oxygen required for the process of biological degradation is extracted from the air and introduced in an energy-conserving way into the mixture of sewage and activated sludge with the aid of jets situated on the floor of the reactor. In such towers, 60-80 per cent of the atmosphere oxygen is used by the micro-organisms during the degradation of organic pollutants instead of only 5-10 per cent in conventional processes. As a result, less waste air is accumulated, which is collected and given secondary treatment in order to prevent noxious odours.

That waste water produced by the Bayer factories which is not used as cooling water but is polluted, flows separately to the joint sewage plant where it is first neutralised by the addition of a lime slurry. In order to remove solid matter, the factory waste water is purified by mechanical means in eight sedimentation basins and then purified again using the tower system. The waste water then flows into two open aeration basins and three secondary sedimentation basins as well as six Dortmund wells in order that the activated sludge can be separated. The purified water is channelled into the River Rhine after passing through a monitoring station, part of the sludge is re-fed to the tower and the remainder is extracted and, following a thickening process, disposed of as sewage sludge.

Source: Bayer AG, Germany.

Figure 19. **ICI deep shaft process achieves high oxygen transfer by using increased hydrostatic pressure**

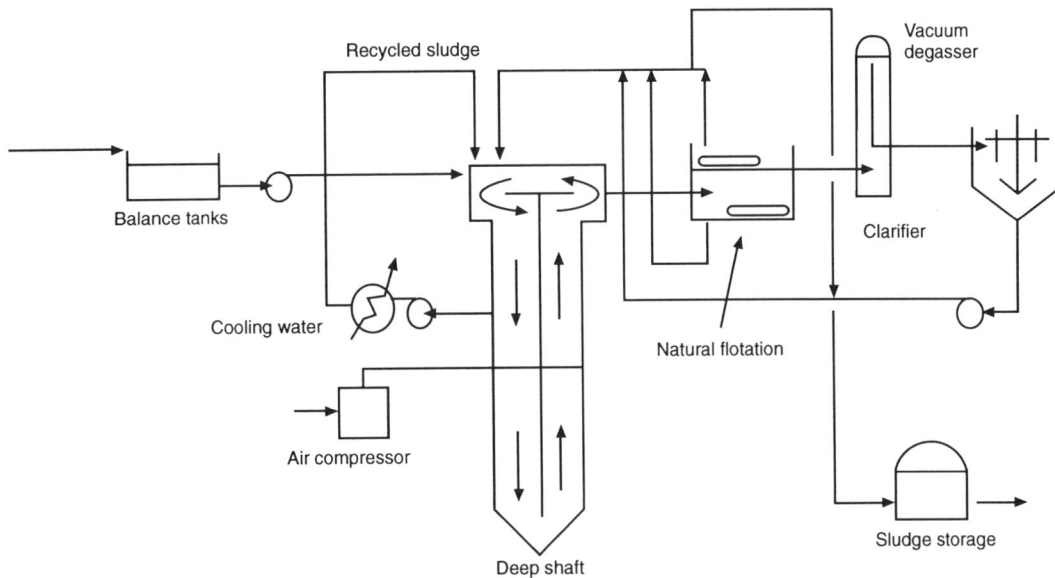

Source: R. Atlas, University of Louisville, United States.

degassed before settlement. The system has low running costs but can be very expensive to build because of the structural costs. The advantage of using oxygen is that much higher levels of biomass can be maintained per unit volume as mass transfer limitations are overcome; as a consequence much greater quantities of BOD per unit volume can be removed. The operating costs are in general greater than using air, although the process can be economical for industrial waste waters where land is limited.

Sludge treatment

Anaerobic digestion is the most common type of sludge treatment and there are several types of anaerobic reactor for this purpose. The choice between them depends on the type of waste to be treated. The organics in domestic refuse are mainly solids (30-40 per cent solids), the organics in domestic sludge and agricultural wastes are slurry (5-8 percent solids) and the pollutants in industrial waste are in solution or colloidal suspension. Industrial effluent is the most amenable to treatment.

Reactors for the treatment of sewage sludge and agricultural slurries are, traditionally, stirred tanks often mixed by methane generation. Retention times are governed by the rates of solids hydrolysis. Rates of solids breakdown are rapid for up to 15 days retention, but then become slower as the easily metabolisable carbohydrates, proteins and fats are exhausted. There are always indigestible fibre residues of between 10 and 30 per cent of the volatile feed solids even at retention times greater than 30 days. Typical retention times for solids and slurry digestion are between 12-20 days and this is sufficient for microbial growth to replace organisms passed out with the treated waste. If there is no system for keeping or reusing bacteria then the minimum retention time in the reactor is limited by microbial growth rate. The doubling time of the methanogens is five days, and this is too long for a commercially viable anaerobic reactor of reasonable size. Different reactor designs which hold back biomass within the reactor or which recycle the bacteria after separation have therefore been developed for industrial waste. Biomass retention is then no longer dependent on liquid retention time and high bacterial concentrations can be obtained.

Appendix 8

Inorganic Pollutants

Heavy metals

Essential metal ions have some important biological functions, and specific regulated systems for their transport into microbes have evolved. Many essential metals are toxic to microbes at high concentrations, and homeostatic mechanisms function in microbial cells to maintain the appropriate balance for effective biological function. The uptake of individual elements by microbes depends on both their chemical and physical properties. Carbon, nitrogen, oxygen and hydrogen are all required in bulk amounts, while requirements of some elements seem to have been determined by their relative abundance in the earth's crust and their solubility in water under anaerobic conditions. Consultation of crustal abundance data for elements indicate that 27 of the 34 elements that occur at concentrations levels in excess of 0.1 µg/l, are essential for microbial growth. Of the four metals that are of greatest concern with respect to adverse affects on human health, only chromium is an essential nutrient. Both lead and cadmium are non-essential for microbial growth, while mercury concentrations are below the threshold for inclusion in the list of crustal abundance.

Toxic metals are present both in suspension and solution in the feeds to many waste water treatment facilities. A portion of the particulate heavy metal content is removed during primary (physico-mechanical) treatment by sedimentation, but metals present either in finely divided particulate form or in a dissolved state enter the biological treatment stage where they can be eliminated, modified or remain unaffected.

The removal of toxic metals during waste water treatment is a process of increasing importance both in the industrially developed and in the newly industrialising countries. Traditionally, the performance of treatment processes has been based on "lumped" parameters describing the concentrations of bulk pollutants. More subtle but highly toxic pollutants, such as metals that endanger human health, have, until relatively recently, been very largely ignored. Biological processes are clearly important in eliminating metal pollutants, but for the degree of removal that will be required for the future, such processes require optimisation.

Toxic metals are also present in contaminated land and in groundwater as a result of historical industrial activity. Continuation of the activity requires that process water and abstracted groundwater must be purified before discharge (Case Study 19).

The heavy metals that are of the greatest concern with respect to human health are mercury, cadmium, chromium, and lead; the WHO guideline values set for these metals are less than 0.001, 0.005, 0.05, and 0.05mg/l, respectively, in drinking water. Heavy metal discharges can result in severe ecotoxicological problems in receiving waters, thereby restricting the effectiveness of natural self-purification mechanisms. While the uptake of toxic metal ions by microbes in conventional activated sludge treatment processes has been an accepted process feature for many years, such mechanisms have been subjected to little more than cursory investigation as far as waste water treatment is concerned. It is also most important to remember that, unlike biodegradable pollutants, sorbed metal ions are generally converted into either toxic products or residues that remain associated with the process microbe/biopolymer matrix and can either be released during sludge treatment or be remobilised after sludge disposal.

Although the biological mechanisms directly associated with metal modification have major implications with respect to the efficacy of biotreatment processes, they have been very largely ignored because of the apparent ability of so-called non-viable activated sludge flocs to remove percentages of heavy metals from solution similar to those removed by so-called viable activated sludge flocs. However, in developing such a concept, the term viability is inadequately defined. Just because a microbial cell is apparently non-viable according to some test procedure, it is not necessarily so and certainly not necessarily devoid of metabolic activity. Because of the suggested lack of linkage between metal uptake and viability, and hence, metabolic activity, it is generally assumed that the biological factors which affect the extent of metal immobilisation by activated sludge flocs are restricted to those which influence the chemical nature of surface binding sites, those

Case Study 19: Shell's SRB process for metal immobilisation from aqueous effluents – the Netherlands

Historically, metal refiners have produced solid wastes and sludges that contain toxic heavy metals and sulphate. These wastes, which have often been stored at processing sites, can be leached by rain and surface water, thereby contaminating the groundwater. This problem is exemplified at the century-old Budelco zinc refining site in the Netherlands where unacceptably high levels of zinc, cadmium and sulphate occur. While abstraction of groundwater with a geohydrological control system will prevent spreading, the heavy metal and sulphate-contaminated aqueous extract must be purified before discharge to the public domain.

In the Shell SRB process, R&D was directed to find a solution that would fit the existing infrastructure at Budelco and produce an aqueous effluent that would be consistent with European legislation. A biological solution could only be successful if it proved to be robust enough to fit into the metals processing environment. After two years of laboratory research, sufficient data had been generated at the one litre scale to warrant construction of a demonstration unit at Budelco. This 9 m³ sludge blanket reactor unit was used to confirm the laboratory process operating window. In addition, it showed that the abstracted groundwater could be treated and that the aqueous process effluent would indeed meet European standards.

Sulphate-reducing bacteria (SRB) convert sulphate to sulphide, precipitating heavy metals as their insoluble sulphides. In nature, these bacteria are part of a consortium of anaerobic organisms each having their preferred simple carbon growth substrate, which is also the energy source for the reduction of sulphate, but all requiring neutral reducing conditions. The microbial process is operated under non-sterile, neutral conditions using ethanol as the growth and energy substrate. Ammonia and phosphate are added as nutrients. Sulphate concentration in the reactor's aqueous effluent is controlled by the ethanol-to-sulphate ratio in the process feed. An oxidation potential lower than –300 mV is needed to maintain maximum methanogen activity so that the small amount of acetate produced (an SRB by-product) is converted into carbon dioxide and methane, ensuring that the effluent has a low biological oxygen demand. A flocculant is added to the reactor to maximise retention of the very small particles of metal sulphides. The effluent, containing soluble sulphides, can be sweetened with an aerobic microbial process before being discharged to the environment and hydrogen sulphide must be removed from the gaseous stream before the methane is flared. The sludge stream, which contains the metal sulphides, can be recycled to the zinc refinery roaster.

Further process development was carried out by Budelco personnel, in collaboration with PAQUES Environmental Technology, the designer and constructor of the commercial plant (see diagram below). The commercial process, which was commissioned in 1992, just six years after initiation of the project, uses an SRB reactor with a volume of 1 800 m³ and is capable of treating 7 000 m³ per day of contaminated groundwater with recovery of the heavy metals and sulphur.

The SRB process can treat contaminated aqueous streams and produces an effluent with environmentally acceptable concentrations of heavy metals and sulphate (ppbs and 100 ppm respectively). The commercial process is easily controlled and needs little supervision. Both capital and operating costs of a commercial process compare very favourably with alternative chemical and physical clean-up processes.

Diagram of SRB pilot plant

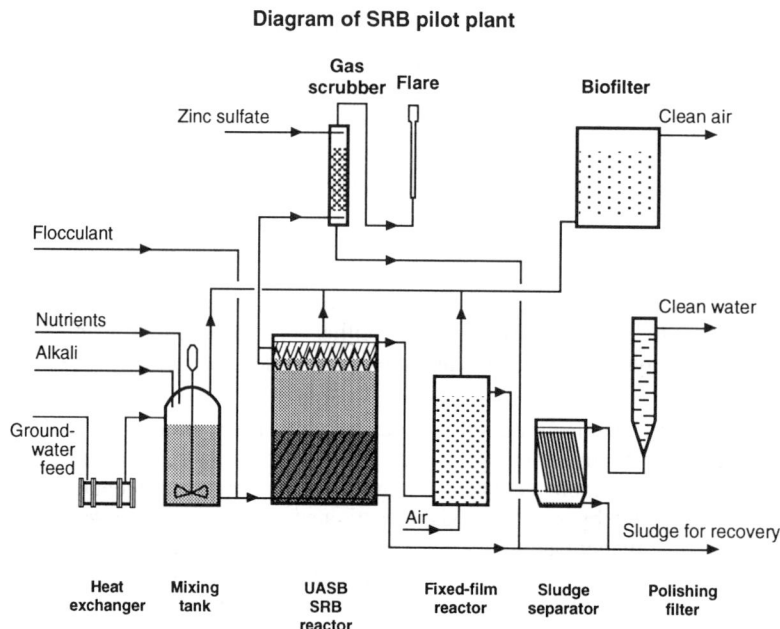

Source: Shell Research Limited, United Kingdom.

112

Case Study 20:   Dentrification of drinking water in a fluidised bed reactor – Blankaart, Belgium

A semi-industrial fluidised bed reactor (40 m³/h) for heterotrophic denitrification using methanol as reductant has been tested for the removal of nitrate from surface water at the Blankaart drinking water production centre.

The process technological aspects and overall performances have been evaluated in a pilot plant which consists of a rectangular tank (see diagram). The reactor was filled with sand to a depth of 2.7m and was operated in upflow mode resulting in a fluidisation and expansion of the sand bed by about 25 per cent. The sand bed expanded more and more as it gradually became covered with biological growth to a maximum of 66 per cent. The height of the bed was intermittently controlled by drawing the coated sand to a centrifugal pump and an in-line mixer which sheared the biomass from the sand. Sand and biomasss were separated, the cleaned sand returning to the reactor and the biomass being stored prior to sludge treatment.

Methanol was provided as a carbon source and in order to limit the formation of nitrite in the reactor it was necessary to work under nitrate-limiting conditions with an excess of methanol in the reactor effluent. Biological nitrate removal is followed by trickling filters in order to reoxygenate the water and to remove the excess methanol.

It was decided that a full scale fluidised bed facility of 750 m³/h capacity, with the following characteristics, would be constructed:

Feed flow rate (m³/h)	750
Reactors - number	2
Surface (m²)	19 each
Height (m)	6.2
Upflow velocity (m/h)	20.00
Sand characteristics (mm)	0.2 - 0.5
Initial sand bed height (m)	3.2
Initial fluidisation height (m)	4.0
Design performance at 2°C (mg/1 nitrate)	75.0 -> 0.0

The methylotrophic denitrification reactors can easily be incorporated into the existing water production plant without major plant expansion. The reactors will be designed so that a maximum of 50 per cent of the total production capacity will be completely denitrified. This provides additional dilution of residual methanol in the reactor effluent. Furthermore the existing downstream processes, including trickling filtration and granular activated carbon filtration, guarantee that no methanol can reach the distribution system preventing the risk of any bacterial aftergrowth during transport of the water to the consumer.

Diagram of nitrogen removal plant

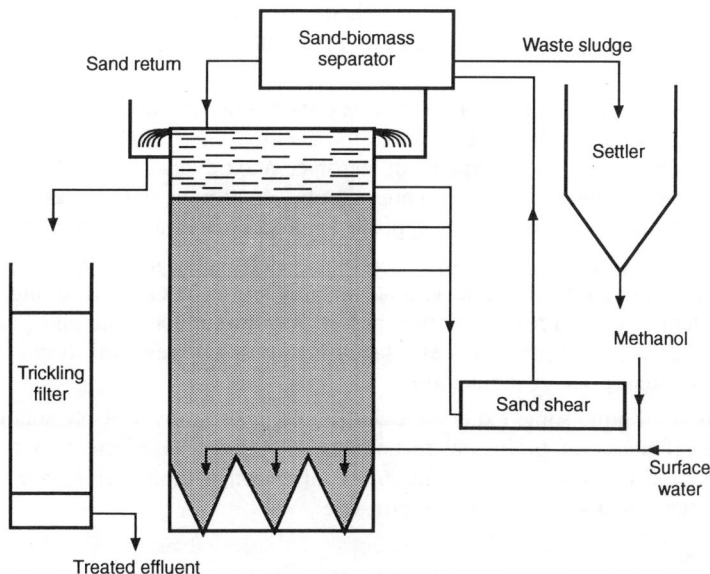

Source: J. Liessens, R. Germonpre, S. Beernaert and W. Verstraete, 1993.

which control the mechanical properties of sludge and those which control the distribution of soluble ligands having a detrimental effect on removal by maintaining metals in solution.

Nitrogen

Levels of nitrogen are of increasing concern, particularly in drinking water, and considerable work has gone into developing processes for its removal. Ammonia is generally oxidised to nitrate (nitrification) and the nitrate is converted to nitrogen gas (denitrification).

Biofilm systems, used in water and waste water treatment, can remove ammonia and nitrate in certain circumstances:

i) Ammonia removal: If the carbonaceous organic loading is less than approximately 0.12 kg BOD/m^3 reactor per day, there are no inhibitory substances present or the temperature is not too low, nitrification will occur. Ammonia removal can also be achieved in separate post-nitrifying filters *i.e.*, treating secondary effluent after the majority of the carbonaceous BOD has been removed.

ii) Denitrification of waste waters: The heterotrophic bacteria in activated sludge are able to utilise nitrate for respiration in the absence of oxygen. In a conventional activated sludge plant at least 50 per cent of the flow is recycled and enriched with nitrate from the oxidation of ammonia and organic nitrogen. The first third usually of a plug flow activated sludge plant is made anoxic or has no free dissolved oxygen, and the high substrate concentration is used to encourage the micro-organisms to utilise nitrate instead of oxygen for their respiration. The transient exposure to anoxic conditions has no adverse effects on the process and the added benefit of removing nitrate as nitrogen gas.

iii) Denitrification of potable waters: Fixed film and fluidised bed systems utilising both heterotrophic and autotrophic micro-organisms have been used for nitrate removal from ground and surface waters for potable water treatment. Heterotrophic systems have to be supplied with a carbon source, usually acetate or ethanol, and can be operated in nitrate-limited or carbon-limited mode. In the former case, almost 100 per cent nitrate removal can be achieved, and only part of the flow will be treated for blending to get below recommended levels. With carbon limitation, all the flow is treated to reduce the nitrate to below the recommended concentration (Case Study 20).

Phosphorus

Phosphate enters water bodies from a variety of point and non-point sources. Municipal waste waters are a controllable point source and removal of phosphorus at waste water treatment plants is desirable to alleviate environmental damage. Chemical removal of phosphates has centred around precipitation techniques using lime, alum and ferric chloride. However, these treatment methods are relatively expensive and produce sludge that is difficult to dispose of. Crystallisation of phosphate onto a particulate support such as sand is now a commercial process in the Netherlands.

Potentially cheaper biological phosphorus removal is possible in activated sludge plants modified to a plug-flow configuration and with an anaerobic zone prior to the aeration tank. This is the simplest configuration and there are several others involving combinations of aerobic, anoxic and anaerobic zones. Under anaerobic conditions, the sludge releases phosphates, while under aerobic conditions phosphates are taken up by microbes in excess of normal metabolic requirements and deposited intracellularly as polyphosphate.

The theory behind the observation is that aerobic bacteria can normally derive some energy from polyphosphate storage compounds while their aerobic metabolism is blocked. In doing so the organisms will become phosphate-deficient so that on reaeration they absorb and store much more phosphate in "luxury uptake" mode. A three-stage activated sludge process is envisaged for phosphate removal. It consists of an anoxic zone, followed by an anaerobic zone prior to the aeration.

Although there has been quite some experience with biological removal of phosphorus, both in the United States and Europe, the efficiency of biological techniques for phosphate elimination have been disputed and considerable controversy exists with respect to this mode of action, although *Acinetobacter* spp. are generally considered to be the bacteria that are primarily responsible.

R&D Priorities and Policy Recommendations
in Environmental Biotechnology

During its third meeting on 11-12 January 1993, the Ad Hoc Group of Government Experts on Biotechnology for a Clean Environment agreed to hold a workshop of scientific experts to identify R&D needs and priorities in environmental biotechnology. This workshop took place at the Oatlands Park Hotel in Weybridge, Surrey, United Kingdom on 3-4 June 1993, and was sponsored by the UK Department of Trade and Industry.

The following text is the result of the workshop discussions. It represents the view of the participating scientific experts (see list attached), and not necessarily those of their countries.

Introduction

In the face of the increasing severity of environmental damage seen during this century, there has been a growing awareness on the part of the public and governments that actions are needed to maintain and restore environmental quality. While waste minimisation and recycling programmes have been instituted in many industries and by public authorities to help conserve resources and to protect the environment against the release of wastes and pollutants, new technologies are needed to aid in sustaining development and environmental quality. The accelerated development of biotechnology during the last decade presents new possibilities for dealing with both the current and emerging problems.

Biotechnology has provided solutions in the past for environmental needs. When the disposal of wastes into rivers threatened human health and the well-being of aquatic life, wastewater treatment facilities were developed during the nineteenth century as one of the first applications of biotechnology for the maintenance and restoration of environmental quality, and this treatment has been of great benefit to humankind. Since that time there have been only minor changes in the fundamental designs of the original sewage treatment plants and the way organisms are used, and there are now a significant number of failures of these facilities to meet performance criteria.

Despite the long-term use of sewage treatment facilities, composting, and landfills, to deal with environmental problems, there has not been a significant breakthrough in the use of biotechnology to deal with pollutants generated by modern twentieth century industry. Individual applications of biotechnology have been used for bioremediation of polluted sites *in situ*, but biotechnology has not yet become an integral part of overall pollution remediation efforts. Site specificity and the inability to predict and to monitor performance have limited the acceptance of biotechnological solutions by engineers and managers charged with the responsibility of deciding on appropriate environmental remediation solutions.

Biotechnology is one of several available technologies for the maintenance of environmental quality and must be viewed within the larger spectrum of scientific and engineering disciplines. However, the importance of biotechnology within this overall context has increased significantly within the past five years and will continue to increase. This is in part due to the exceptionally rapid advances of knowledge, the low environmental impact and cost effectiveness of using biological as opposed to physico-chemical treatments and also to problems associated with non-biological treatments such as, for example, the generation of gaseous pollutants during the destruction of others by incineration.

Emerging threats to the environment, such as the atmospheric changes that potentially pose a threat of global climate change, the loss of trees and forests due to atmospheric pollutants, and the formation of deserts, are appropriate topics for biotechnology. Biopolymers, synthetic fuels and other biological alternatives to chemical processes present opportunities for economic and environmental benefit.

This New Industrial Technology System must harmonise with the global material cycle in the biosphere. It should be gentle to humans and to the environment, and have four key characteristics: the renewability of resources; the mildness of production processes; the compatibility of products with humans and the environment; and the recyclability of wastes. We expect that biotechnology will play a major role in this transformation to a new system.

Traditional and mainstream wastes

The major problems of solid and liquid waste had already arisen by the nineteenth century and continue today as a result of the development of large urban areas and the necessity of multiple uses of land and water both as receivers of wastes and sources of potable water. The treatment of wastes from domestic and industrial sources, in facilities such as activated sludge and composting plants, represents the mainstream of environmental biotechnology. Biotechnological processes will continue to play a central role in the treatment and recycling of municipal, industrial, and agricultural wastes. Shortcomings in perception, policy and management practices in the past have led to accumulating burdens – such as the chronic pollution of water, the massive amounts of surplus sewage sludge, excessive ammonia emission from livestock production, and poorly managed landfills – which are threatening our future.

Nowadays, environmental biotechnology can avoid adding to these problems, and should contribute better solutions and management systems for the treatment of these traditional wastes. To achieve sustainability in consumption and production processes and improve the quality of bodies of water or soil receiving the residues, significant reduction in the emission of waste products at each step in the cycles of production and consumption is required. This can be done by developing new, comprehensive waste management systems that emphasise recycling at each step. These unit processes must be integrated into production systems, taking into account the scope and limits of biotechnological processes.

Although environmental biotechnology is one of several approaches to waste treatment, it is often the technology of choice because it is basically an acceleration of natural cycling processes and it can be achieved with minimum costs and often maximum recuperation. The process of waste treatment may be much improved, however, particularly with respect to the major by-products, which comprise:

- noxious off-gas emissions to air, which must be avoided;
- surplus sludge production, which must be minimised;
- effluent by-products for which new uses must be found;
- effluent water quality must be improved so that re-use becomes possible.

To achieve improvements in waste treatment systems, new combinations of physical/chemical and biological (including both anaerobic and aerobic) treatment processes must be developed, taking into consideration the interfaces between them.

A. Current inputs of pollutants into the environment are diverse and variable. New chemicals are constantly introduced. Substances which may be spread in water, waste and soils, should be restricted to those which proven tests have demonstrated to be manageable. The design of well-defined officially approved test procedures for better assessing substances which are compatible with available disposal technologies is needed. This should allow public authorities to implement appropriate counter-measures utilising biotechnology.

Research should focus on the development of well-defined reliable test procedures to assess the fate of anthropogenic chemicals in full-scale treatment systems and in ecosystems. For a number of pollutants, (bio)sensors need to be developed.

B. Biological processes in drinking water production, waste water treatment, solid waste conversion and off-gas treatment only operate optimally for perhaps two-thirds of the time. They are essentially "black-box" operations and lack reliability and predictability. The black box must be analysed to reveal individual components and processes that can be understood, optimised and controlled.

Research should focus on the understanding of these individual processes and on the development of measurement systems and models for their control.

C. Most available and foreseeable technologies for waste treatment plants rely on the biological activities of microbial communities. The performance and predictability of these communities is insufficiently understood.

Study of the organisms and their interaction (microbial ecology) in the microbial communities involved in environmental processes is needed. The utility of genetically modified organisms (GMOs) with novel biodegradative phenotypes must be assessed. Inocula with known composition and functionality should be developed.

D. Waste treatment processes generally deal with complex wastes and are insufficiently directed at specific recovery of products. Production of value-added products, *e.g.* waste water treatment systems producing directly reusable water, will become essential.

Research should focus on the design of dedicated waste treatment processes aimed at specific wastes and the recovery of useful (value-added) products.

E. Many wastes contain large quantities of nutrients and valuable humic materials. The sites of waste treatment are generally distant from locations where these nutrients can be usefully applied.

Research should focus on the reuse of nutrients and other end-products from environmental processes by developing practical standards for recycling. Investigations should be performed on the wider use of recovered products (locally and internationally) by demonstrating their value and safe reuse.

Contemporary industrial contaminants

Important sources of contamination, particularly during the latter half of the twentieth century have been releases of petroleum hydrocarbons and industrial chemicals, especially those containing chlorine, that have polluted both soils and waters. Localised sites have been contaminated, some of sizeable area. These are the concern of individual nations. In the treatment of aqueous and gaseous effluents, biotechnological processes are established technologies based on a history of traditional performance. There is, however, a lack of experience in the treatment of soil, sediments and groundwater which has created some scepticism concerning the application of biotechnology in this specific area. It is an article of faith in microbial biochemistry and physiology that the majority of pollutants are biodegradable, that is can be mineralised or modified by some obtainable microbial population. However, some compounds that are biologically attacked are not completely degraded, for example, some pesticides. In such cases, the original compound may disappear but other compounds, resistant to further degradation, may accumulate.

The majority of current site-specific problems of contamination and pollution can be treated by microbiological processes based on indigenous organisms. For the adequate treatment of some pollutants, however, it may be necessary to exploit genetically modified organisms (GMOs) specifically designed for the problem in hand. For the most part, GMOs that may be considered for this treatment will have altered genetic regulatory functions so that they overproduce enzymes that they would naturally produce in lesser quantities.

The key areas for the decision-making process in the treatment of industrial contaminants that have been released into the environment at a specific location are:

- The chemical definition, as far as possible, of the contaminants: what are their concentrations and to what extent must these be reduced to ensure ecological and human health?
- Are the contaminants inherently biodegradable?
- Are there limitations to the removal of the contaminant and, in particular, is there suitable technology to solve the problem at a contaminated site?

These questions have to be answered in a quantitative manner to allow the comparison with competing technologies (physical/chemical etc.).

A. The decision to implement a specific remediation option should be based upon information that allows the best alternative to be identified. In the absence of other overriding considerations, a proper site assessment is crucial not only for decision making but also to provide the baseline data essential for monitoring process efficiency. Site assessment should provide a variety of physico-chemical and biological information, including the bioavailability of target compounds, catabolic potential of the indigenous microbial community, and the physico-chemical requirements of these activities. Though thorough site assessment is at present time-consuming and costly, it will become more rapid and less expensive as experience is gained. This experience is essential to develop generic strategies for remediation.

The measurement tools, procedures and protocols for site assessments and monitoring should be improved. The relationship between laboratory studies and field performance of technology (scale-up) should be established through the use of model ecosystems (microcosms).

Contaminant availability in environmental matrices should be investigated. Toxicity tests able to operate "on line" to assess when a soil is contaminated or decontaminated should be developed. Generally acceptable clean-up criteria need to be formulated.

B. Bioremediation processes generally involve microbial consortia as catalysts. The efficacy of such processes largely depends on the activities of and interactions between different species in consortia which are presently poorly characterised.

The simplest and preferred process of biocatalyst development involves indigenous organisms which can be stimulated *in situ* (*e.g.* through the addition of carbon, nitrogen, phosphorus, oxygen, etc.) to obtain a satisfactory level of remediatory activity. Alternatively, organisms possessing the desired biological potential can be added to the site. These may be minor site organisms possessing the desired activity, and/or engineered organisms not derived from the site.

The biochemistry, genetics and regulation of catabolic pathways require more intensive investigation. This should include improved selection, enrichment and adaptation strategies for organisms with improved catabolic activities. The knowledge obtained could be exploited to design higher performance biological catalysts and to optimise microbial efficiency in in situ and in bioreactor-based processes.

Research to optimise microbial performance both in-situ and in bioreactors should be strengthened.

Research should be conducted into microbial ecology, particularly microbial community structure and population dynamics in response to environmental stimuli, and to the activity and survival of induced/introduced populations.

The construction and performance of designed microbial consortia to meet the remediation of waste streams containing industrial contaminants should be explored.

C. Biotechnological processes cover "intensive" systems where contaminated material is subjected to a closed "reactor" environment and "extensive" systems where biological catalysts are used in the wider environment. The decision to apply a biotechnological approach will have required an engineered design with predictable time scales, costs and final outcome that can be monitored and controlled. This is more readily achievable in the treatment of gaseous and aqueous effluents than in the remediation of soil and groundwater.

Wider knowledge is needed both on the biological component of these processes and also on the specific nature of the environmental compartment and the engineering technology applied.

Laboratory experimental systems need to be developed to provide field engineering design parameters and performance prediction. The movement of nutrients and organisms through water-saturated and non-saturated soil matrices must be investigated in order to provide engineered design parameters for technology application.

The efficacy, rates and economics of biotechnological solutions for site remediation technology should be assessed by the use of demonstration sites with necessary controls.

Emerging global environmental problems

Pollutants that are widely spread, primarily through the atmosphere, are an emerging environmental problem as also is the depletion of scarce resources. Preventative measures need to be implemented and biologically synthesised materials developed to replace those chemicals that have caused global environmental deterioration. These problems transcend national boundaries. The massive use of fossil fuels has led to increased emission of carbon dioxide. Industrial activity has led to production of other greenhouse and ozone-depleting gases such as CFCs (chlorofluorocarbons such as Freon). Increasing amounts of methane are being released as a result of rice production and cattle breeding. Such global changes in the atmosphere may emerge as major problems of the next century unless preventative measures are implemented.

The twenty-first century will see the possibilities of applying biotechnology to global, in addition to local, environmental threats: three have been identified which might be resolved by biotechnology and specific research targets recommended. These are: atmospheric quality (greenhouse and acid rain gases); depletion of natural resources (fuels, materials); water shortage and desertification.

A. Atmospheric carbon dioxide may be removed or "fixed" by green plants, algae and bacteria. Some fossil fuels may be replaced by renewable fuels produced by plants or micro-organisms.

A fundamental understanding is required of plant and microbial genes and their translation into activity in order to permit the development of quick-growing trees, plants for fuel, algae and micro-organisms for CO_2 fixation and hydrogen production.

Studies of the relationship between structure and function at the enzyme level and at the level of the metabolic pathways are essential to optimise treatment processes such as fixation of CO_2, nitrogen and sulphur compounds, and the production of biopolymers and renewable fuels.

Research should be undertaken into the selection of plants, algae and bacteria for improved fixation. This should include, for example, selection of coral reef organisms and synthetic support structures for reefs. Research is also required into autotrophic micro-organisms and their activity on support media to permit the use of biofilters for treatment of point sources of CO_2 emission.

Research should be supported on organisms that produce ethanol and other potential fuels at high concentrations. The mechanisms for the production of hydrogen and the selection of superior organisms should be investigated; the scale-up of this process will require extensive investigation.

B. Acid rain has many causes, particularly the emission of sulphur and nitrogen gases from power plants, automobiles, animal housing, and waste water treatment.

Biofilters should be developed for the removal of organohalogens, sulphur and nitrogen compounds from stationary point sources.

A fundamental understanding of the natural ecosystems of forests should be sought with a view to adjusting them and thus protecting the forests. Research into the interaction of organisms, plants and their environment, in selected ecosystems, for example paddy fields, is required in order to modify them to achieve specific ends such as reduced methane emissions, etc.

C. Natural resources are being depleted at an increasing rate: forests, water, fossil fuels, metals, etc. Biotechnology may make renewable resources more economical and aid resource recycling. Bioprocesses may increasingly replace conventional processes, especially in the chemical industry.

Investigation into organisms producing polymer precursors is required. Replacement where possible of materials derived from fossil fuels by renewable materials should be encouraged.

D. Metals are released into the environment from industrial and domestic sources as well as following waste treatment. These metals cannot be destroyed by physical or biological treatments.

Biological processes must be developed to recover metals and make them available for recycling.

E. The large-scale use of fossil water, changing climate and increased water consumption as a result of demographic changes are putting ever heavier pressure on the water resource. Also, partly as a consequence of human activity, deserts are increasing in area in every continent. Water can be made more available by aiding the cleansing of large bodies of water.

Research is required into the mechanisms by which plants survive in conditions of low water and brackish water with the aim of developing superior plants for these conditions. Super-absorbent biopolymers require research as water-retaining materials in desert conditions. Research is required into the selection and development of plants and micro-organisms, with an understanding of the ecosystems, which will improve the quality of lakes, etc., by removing inorganic pollutants (nitrogen, phosphorus, metals, etc.)

List of Participants

Workshop on Weybridge, United Kingdom, 3-4 June 1993

Chairman
Dr. Ronald ATLAS
Department of Biology
University of Louisville
KY, **United States**

Dr. R. BOLLIGER
EBIOX AG
Sursee, **Switzerland**

Dr. Wulf CRUEGER
Bayer AG
Env. Protection Dept.
Leverkusen, **Germany**

Dr. Victor DE LORENZO
Centro de Investigaciones Biologicas
Madrid, **Spain**

Dr. Thomas EGLI
EAWAG
Dübendorf, **Switzerland**

Dr. W. HARDER
TNO Institute for Environment Sciences
Delft, **The Netherlands**

Dr. Toyohiko HAYAKAWA
Chiba Institute of Technology
Narashino-Shi, **Japan**

Dr. D.B. JANSSEN
Dept. of Biochemistry
University of Groningen
Groningen, **The Netherlands**

Dr. Ebba LUND
The Royal Veterinary and Agricultural University
Frederiksberg C, **Denmark**

Dr. Andrea ROBERTIELLO
Eniricerche Laboratory of Microbial Ecology
Monterotondo (Rome), **Italy**

Dr. Ichiro SHIMIZU
Japan Bioindustry Association
Tokyo, **Japan**

Dr. K. TIMMIS
Gesellschaft für Biotechnologische
    Forschung (GBF)
Braunschweig, **Germany** (EC)

Dr. Willy VERSTRAETE
Environmental Protection Department
Ghent, **Belgium**

Dr. R. WATKINSON
Shell Research Limited
Sittingbourne Research
Sittingbourne
Kent, **United Kingdom**

Dr. Elisabeth A.M. BAKER,
Head of Biotechnology Branch,
Chemicals and Biotechnology Division
Department of Trade and Industry
London, **United Kingdom**

Dr. M.H. GRIFFITHS
Chairman, OECD Ad Hoc Group of
    Government Experts on Biotechnology
    for a Clean Environment,
Mike Griffiths Associates
Surrey, **United Kingdom**

Dr. Peter J. NOLAN
Head of Biotechnology Unit
Department for Trade and Industry
Chemicals & Biotechnology Division
Laboratory of the Government Chemist
Middlesex, **United Kingdom**

Dr. Julie RANKIN
Department of the Environment
London, **United Kingdom**

OECD Secretariat

Dr. Salomon WALD
Dr. Yoshitaka ANDO
Mr. Mark CANTLEY

Part II

ECONOMIC AND INDUSTRIAL ASPECTS

Chapter 6

An Economic Perspective on Environmental Biotechnology

Part I of this report reviewed the scientific and technological state of the art in environmental biotechnology. Part II will analyse economic and industrial aspects of this technology, particularly the strategies of industrial companies. The first chapter attempts to outline the broader economic framework in which industrial companies will be likely to operate and to develop some rather speculative ideas about long-term trends.

This chapter refers repeatedly to environmental policies because they are a key determinant of the economic framework of environmental biotechnology. It must be clearly understood, however, that this chapter, and indeed the entire report, is not about environmental policy as such and does not make environmental policy recommendations.

1. From science push to market pull

In the early 1970s, three sectors were expected to become, virtually simultaneously, the main fields of application of new biotechnology: human health, agro-food production and the protection and restoration of the environment. While these predictions have been realised for health and, to a lesser extent, due to difficulties and delays, for the agro-food sector as well, it has taken more than 20 years for recent biotechnology to begin to have a noticeable impact on environmental protection. This is somewhat surprising, considering that one of the earliest and widest applications of traditional biotechnology was the treatment of waste water in facilities built in the nineteenth century to safeguard public health in urban areas and reduce water pollution. In all countries their introduction took place without opposition, or even much notice, from public opinion. Thus, one might have expected a more continuous stream of new or improved biological applications to environmental problems to take advantage in the advances of the underlying sciences and technologies.

The reason why "biotechnology for a clean environment" did not, until recently, show the same rapid development that occurred in the medical, and to lesser degree in the agro-food sectors, has much to do with the "science-push" character of modern biotechnology in general, at least during its early years. "Science push" is a term used by economists for technological developments which are driven largely by scientific research initiatives. More than any other major technology of the twentieth century, the technology which harnesses the power of living organisms and puts them to good use was born in universities and was nurtured by scientists with traditional scientific motivations. Its trajectory was first determined by little other than scientific competition and the excitement which the prospect of great discoveries creates in the human mind.

One problem of environmental biotechnology was that it often could not compete in glamour with medical or agricultural biotechnology. Scientists, students, and research funds were more easily attracted to the new molecular biology which promised to cure cancer or to engineer high-yield rice, than to the skills necessary to improve, for example, sewage disposal. Past differences in priorities are still clearly visible in today's funding patterns: the US federal government budget, for example, planned to spend $83.3 million for environmental biotechnology R&D in 1993. This represents just 2.1 per cent of federal funds spent for all biotechnology R&D, as against 42 per cent for health and 5.1 per cent for agriculture.

Another important reason why environmental biotechnology has been at a scientific disadvantage is that it does not have the same "natural" R&D constituency as the medical and agricultural research sectors; it is too vast a field, complex and ill-defined. A third reason lies in the greater difficulty of some underlying scientific questions, for example those related to the multitude of interactions between different plant and microbial species in the environment.

During the 1980s, however, the relative scientific disadvantage of environmental biotechnology started to be diminished, if not reversed, by a new development: the introduction, in an increasing number of countries, of environmental protection legislation (clean water, air, etc.), setting standards and forcing manufacturers to comply with them. Thus, the originally weak "science-push" was slowly complemented and reinforced by a new "demand pull" which did not exist in any other sector of biotechnology in such articulated form. There is, of course, a demand pull for new medical discoveries and products (AIDS, cancer, new diagnostics), which gives health biotechnology R&D a strong boost, but this is not comparable with concerted government action to create a new and possibly large market for environmental or environmentally compatible products and processes.

A direct or indirect government defence market for specific products and processes, often coupled with large government R&D expenditures, has been the foundation stone of many major technologies which have changed the face of the twentieth century: air transportation, nuclear energy, new materials, and particularly information technologies. The latter, which have led to the most far-reaching socio-economic transformations brought about by any modern technology, owe their inception (the first computer was for the Manhattan project) and key technological advances (*e.g.* microchips) almost entirely to US defence R&D and to government markets.

Obviously, a technology which has the first step of its R&D chain supported by public money has an enormous advantage over a technology which has to rely on private financing. The latter is applied to the last steps in the R&D chain, rarely to the first.

Environmental legislation as such does not determine the provision of initial R&D money, but it has the potential to create and sustain a market for biotechnology, depending of course, on the latter's technical efficacy, reliability and cost competitiveness. For the first time, some of the macro economic and policy conditions for new biotechnology appear to resemble those on which the earlier success of the other major technologies depended.

2. Government measures and legislation as a new market pull and a "pacer" for biotechnology

Governments and public authorities, responding to a public opinion strongly in favour of "green" policies and technologies, play the leading role in shaping the demand for environmental biotechnology products. Firstly, it is governments which determine whether there will be a market for environmental products at all, by means of legislation and compliance policies. Secondly, apart from the framework conditions, governments are themselves important buyers of environmental biotechnology products to provide for facilities which are usually in the public domain (drinking water, sewage, and waste disposal). Thirdly, government funding of environmental biotechnology R&D has also had some effects on technology trends. In the longer term, financial support for innovative R&D as well as for demonstration of developed processes will become more important. This support, in conjunction with appropriate legislation and other government policies, such as fiscal policies offering tax deductions for specific investments, will determine the direction and pace of technical progress in environmental biotechnology.

The dominant role of government and legislation emerges in all environmental sectors:

Air/off-gas: Government dictates the need to install pollution control equipment, specifies emission standards, monitors their enforcement and affects the economics of pollution abatement control. In some countries, financing for pollution abatement hardware is provided by the government (*e.g.* France) and special tax-based pollution abatement strategies have been developed.

Waste water treatment: Future growth in this segment will primarily depend on the adoption of adequate legislation requiring mandatory water treatment. The amount and types of technology developed and used can largely be determined by government, through input costs (*e.g.* electricity, natural gas, land) and by regulations for the disposal of residual sludges.

Demand, product performance standards, and financing are often dramatically affected by public authorities. For example, a sizeable market for industrial waste water treatment plants has emerged in the Netherlands since the authorities increased the price individual companies have to pay for treatment of their waste water in municipal units. This is also an example of the paramount importance of a domestic market for local companies to gain solid experience for future export activities.

Soil treatment: Government is often the driving force in creating the demand (*e.g.* by obliging the polluter to clean up and dispose of waste), in setting the standards to be met, in regulating the mode of operation of the industry (*e.g.* restrictions on waste transfer from one site to another), and in granting licences to clean-up companies.

Government provides the financing of site clean-up when the original polluter cannot be identified and charged, and it often determines the techniques used.

Solid waste disposal: Government largely rules the industry by granting operating licences to the various industrial plants, regulating the building and operation of waste disposal plants or sites, and specifying the techniques to be applied for waste disposal (*e.g.* by banning landfilling and dumping at sea or by imposing recycling taxes, for example for packaging material). This role of the government in the waste disposal industry is recognised in the United States as the single most important determinant of the industry.

Recent research has emphasised the role of the public sector as a "pacer" in technological innovation (see *National Systems of Innovation*, B.-Å. Lundvall, ed., London, 1992, and particularly Part II, Chapter 7, "The Public Sector as a Pacer in National Systems of Innovation", by B. Gregersen) which may be compared to the role played by a pacer in a cycle race. Most interactions between the public and private sectors result from the public sector's double role as user of innovations and as regulator. Through this double role, the public sector may greatly stimulate or restrain innovation, particularly in a period of technological and market uncertainty.

Carrying the comparison with a bicycle race even further, it has also been said that the pacer (the public sector) will lose contact with the system of innovation if he races too far ahead and that he must himself maintain a high level of scientific-technological competence if he wants close interaction with the private sector in order to encourage innovation. Although a variety of examples can be shown to illustrate this model, one of the most convincing illustrations is environmental technology, and particularly biotechnology.

Legislation and regulations are of little use for environmental biotechnology if there is no enforcement and control. In a number of countries, legislation has been found to be inconsistent, and proper enforcement is lacking in general, or in parts of the country, or in specific industrial, agricultural and energy sectors.

Insufficient co-operation between various branches of government is causing delay and renders the development of an efficient environmental clean-up industry more difficult. Biotechnology companies which are relative newcomers could particularly suffer from legal and regulatory inconsistencies. Also, unresolved political debates, for example on whether incineration or composting should be given priority, or on the minimum level of soil rehabilitation that has to be observed, create great uncertainties for biotechnology companies.

Better co-ordination of legislation can be important at the international level as well, although geographic and climatic differences among countries have to be taken into account. The European Commission, for example, is continuously initiating new legislation, and there are already numerous EC directives on environmental quality in various stages of implementation, with a large potential to affect the search for, and installation of, new technology. Thus, environmental quality standards are no longer the sole responsibility of national governments; some are also set by international organisations.

In spite of the still numerous shortcomings in legislation or implementation, both at national and international levels, the laws that do exist have already created a competitive and expanding environmental biotechnology industry in several countries. As indicated in Chapters 7-12 "Industry Strategies and Perspectives", there is, in countries such as Denmark, Germany and the Netherlands, an obvious causal link between early environmental legislation – which is more stringent in these than in many other countries – and the creation of an environmental biotechnology industry. Germany and the Netherlands are successfully manufacturing and exporting both biological waste water and air treatment technologies, Denmark waste water treatment technologies, etc. Other countries, such as Japan, are making efforts to establish an export-oriented environmental biotechnology industry, and others again have recently introduced more stringent environmental legislation which could, in time, lead to a more efficient environmental biotechnology industry.

The future market potential for such industries could be large, particularly if national and international environmental legislation becomes better harmonised and more strictly enforced. Table 7 gives an idea of the present macro economic size of environmental markets and shows the public and private environmental expenditures as a percentage of GDP in 1990, and in comparison to health, public education and R&D expenditures. Environmental expenditures include all environmental technologies. The part of biotechnology is difficult to calculate precisely, but it could be approximately 15 to 25 per cent (considering that waste water treatment constitutes 30-50 per cent of all environmental treatment, and that about two-thirds of waste water treatment is biological). The share of biotechnology has been increasing since the late 1980s, and will continue to do so.

In comparing expenditures for the environment to those for health, public education, and R&D, one must bear in mind that environmental protection has become a socio-economic objective much more recently than any of the other three. Consequently, expenditure figures for the environment, while still smaller than those for the other three objectives, are rising much faster. Generally they are increasingly by 4 to 6 per cent annually whereas, in most cases, increases for health, education, and R&D are smaller, stagnating or even negative. If these trends remain unchanged, environment expenditures could, in a few years, reach and exceed R&D expenditures and approach expenditures for education. The part of biotechnology in the expenditures for each of these objectives is likely to be largest in the environment sector (15-25 per cent), and smaller in the health and R&D sectors.

Table 7. **Environmental expenditures, compared to expenditures for health, public education and R&D as percentage of GDP in 1990**

	Total environment expenditures [1]	Total health expenditures	Public expenditures for public education	Total R&D expenditures
Germany	2.3	8.1	4.0	2.88
Netherlands	1.8	8.0	6.8	2.17
Switzerland	1.8	7.7	4.7	2.86
Austria	1.7	8.4	5.6 [2]	1.40
United States	1.7	12.4	. .	2.82
France	1.6	8.9	4.7	2.34
Belgium	1.5	7.5	4.9 [2]	1.70
Japan	1.4	6.5	4.6 [3]	2.98 [4]
United Kingdom	1.3	6.2	4.7	2.25
Sweden	1.3	8.8	6.6	2.76 [5]
Italy	1.0	7.7	. .	1.25
Spain	0.9	6.6	. .	0.75

1. Estimates of public and private expenditures, including levies, environmental technology sales, R&D, etc. There is a very small amount of double-counting between environmental and R&D expenditures (Col. 4).
2. Private education included.
3. Public and private expenditures for public education.
4. Over-estimation.
5. Under-estimation.

Sources: Environment figures from *Milieumarkt 7 (3)*, Netherlands, 1993, based, among others, on OECD Statistics (*The OECD Environment Industry: Situation Prospects and Government Policies*, Paris, 1992; and *Pollution Abatement and Control Expenditures in OECD Countries*, OECD Environment Monographs No. 75, Paris, 1993). Health, education and R&D figures from OECD statistics.

3.   Conditions for the diffusion of new technology in general

The preceding discussion has been general. For a more detailed and dynamic assessment, focusing on trends and evolutions, it is useful to draw upon earlier economic impact assessments of new technologies, particularly regarding the conditions and time frame for the diffusion of new technology through the economy. One of the main authors in the economics of technology, Christopher Freeman, has attempted such an assessment for biotechnology in general, in "Biotechnology – Economic and Wider Impacts", OECD, Paris, 1989. One key methodology is to draw parallels with the diffusion of other technologies, although it is very important to recognise not only the similarities, but also the differences between the various generic technologies which have transformed industrial societies over the last hundred years.

In the case of biotechnology, parallels were drawn with the diffusion of the electronic computer and of electric power. The most important lesson that can be learnt from the introduction and diffusion of revolutionary new technologies is that the changes in capital stock, skill profiles, industrial structure, and social organisation which they require are a matter of decades rather than years. Recognition of the relatively long time-scales involved helps to avoid errors both of technological optimism, which tend to ignore some hard economic realities, and of conservatism, which fails to recognise the enormous long-term potential of pervasive technologies.

There is, in fact, a major difference between the diffusion process for a single, even radically different, product, which can be very fast, and the diffusion process of a generic technology with numerous applications in a vast number of economic sectors, leading to a new "technological paradigm" and new technological common sense for a generation of engineers, managers and consumers.

To achieve a major change in a technological paradigm, five conditions should be fulfilled:

i) a new range of technically improved products and processes;
ii) cost reductions for many of these;
iii) social and political acceptability;
iv) environmental acceptability;
v) pervasive effects throughout the economic system.

Information technology satisfied all five criteria, and received, beyond mere "political acceptability", a major push through active government R&D support and markets. In 1985 and after, when comparisons were first attempted, the case for biotechnology in general was not yet clear-cut. It satisfied at least partly the first, but not fully the second criterion, as comparative costs of biotechnology production remained high in certain sectors (*e.g.* health). Also, the problems of social and environmental acceptance encountered by some applications of biotechnology were quite obvious. Regarding the fifth condition, biotechnology could be said to be more pervasive than many other technologies, as it had already a long history with many traditional applications and had also found many new applications in industry, agriculture and various service sectors, but it was still less pervasive than information technology which had penetrated virtually all products and processes of human activity.

Consequently, it was predicted that biotechnology would not become a predominant technology for many sectors and processes in this century. The time scale for biotechnology in general to become a major base for GDP growth, employment, and investment was not likely to be much shorter than the time-scale of earlier pervasive technologies which also depended upon far-reaching structural adjustments, and this could mean a time scale of one generation, or approximately thirty years. In other words, biotechnology could be expected to begin to have major macro-economic effects from approximately 2010 on, although the economic, social and qualitative effects in individual sectors could become important much earlier.

4.   Critical diffusion factors for environmental biotechnology

It is useful to return to the earlier economic analysis of biotechnology, to confront it with new economic data which have been gathered since, and to compare the clearly different conditions in various sectors of biotechnology. Do the general conclusions of a few years ago change in the light of the recent evolution of environmental biotechnology?

A new range of technically improved products and processes

Industry experts state unanimously that the number of environmental biotechnology products and services (mainly bioremediation, but also some prevention and detection technologies) has increased quickly during the last five years (1988-93). In spite of a late start, this branch of biotechnology begins now to resemble the other branches, at least when it comes to numbers.

However, statistical quantification is presently hard to come by because environmental biotechnology is not clearly categorised and can rarely be separated from other biotechnology, or from general environmental technologies of which, in large-scale applications, it forms a part. Patent statistics measure inventions in the environmental area in general, and are not distinguished by underlying scientific disciplines (physics, chemistry, biology, engineering, etc.). It is also clear that the number of industrial companies (both small and large) that are interested or involved in environmental biotechnology has been increasing sharply during the last years, and that technical competence in this field is spreading. Again, no comprehensive statistics are available.

Looking at the criterion of technical improvement over older technologies, environmental biotechnology still suffers from an ''Achilles heel'' which most medical and agro-food biotechnology overcame years ago. Too much of environmental biotechnoogy is still a ''black box'', as has been explained in other parts of this report; the working of biological processes, particularly in bioremediation, is not well understood scientifically, and the feasibility, reliability, repeatability and predictability of these methods are not always sufficiently assured. While this presently puts biotechnology at an industrial disadvantage with regard to more traditional environmental technologies, this will not necessarily remain so.

It could be reversed by a concerted R&D effort to open up the ''black boxes'', which has been proposed in this report as one of the main R&D priorities in environmental biotechnology. Even if such an effort is undertaken, however, a complete change in the present technical disadvantage of environmental biotechnology should not be expected in much less than ten years. Normal time lags between fundamental research and successful technological application of new discoveries in sectors as complex as this one are rarely shorter.

Cost reductions

One of the most critical assets of biotechnology compared to other environmental technologies could be its clear, sometimes massive, cost advantage in virtually every field of environmental remediation. This also sets environmental biotechnology apart from biotechnology applications in other sectors, some of which have proven to be quite expensive (*e.g.* some pharmaceutical products).

Indications that bioremediation might be cheaper than other methods have emerged from various sources during the last years. However, only recently has a more systematic and comprehensive cost comparison been

Table 8. **Costs of biological and non-biological treatment of environmental pollution (soil, air, water) in the Netherlands**

(in Dutch guilders [Gld])

Soil remediation	Gld/ton	Air treatment	Gld/1 000 m³	Water treatment	Gld/m³
In situ:					
Bioremediation (BT)	70-150	Biofiltration (BT)	0.50-5.00	Biological water treatment (BT)	0.10-1
Extraction	125-150	Bioscrubbing (BT)	3.00-6.00	Sedimentation	0.05-30
Electro-reclamation	150-300	Chemical scrubbing	1.00->20.00	Flotation	0.10-2
Steam-stripping	250-300	Adsorption (activated carbon)	1.00-10.00	Adsorption	1.00-10
		Incineration	2.50-25.00	Chemical oxidation	0.50->5
		Catalytic treatment	2.50-20.00	Ultra-filtration	<1.00->20
On/off site:					
Land-farming (BT)	50-140				
Extraction	120-240				
Thermal treatment	100-300				

Note: BT = Biotechnology

Source: This table summarises and compiles figures presented in Tables 15, 17 and 19 of Chapter 7, "Industry Strategies and Perspectives in Environmental Biotechnology".

provided, for the Netherlands. Table 8 compares costs of some biological and some non-biological treatments of soil, air and water in the Netherlands in 1993, showing that biotechnology could in fact be the cheapest method for each sector of the environment. While the exact figures are valid only for the Netherlands, they can be assumed to be generally relevant for all highly industrialised European countries. The same calculations for these countries would probably lead to the same general conclusion. In the United States, similar calculations could, however, yield somewhat different results due to the much lower oil and energy prices than in Europe. For example, thermal methods to clean the environment could be more competitive in the United States than in Europe, although recent case studies indicate that *in situ* bioremediation is still cheaper than other *in situ* methods in the United States.

It is also important to consider that the biological clean-up methods are much more "sustainable" than any others; biotechnology hardly requires raw materials or energy and hardly produces secondary wastes, as pollutants are mineralised into harmless natural substances such as H_2O, CO_2 and $NaCl$. In contrast, all other technologies require chemicals and/or energy, and wastes are often only shifted to other environmental compartments (*e.g.* from water to air). Of course, the "sustainability" advantage is partly captured in the cost calculations.

However, when one compares the costs of competing technologies, one has to bear in mind that a new technology needs to show major, and not just marginal, cost advantages if it wants to replace an existing installed technology, because of the latter's fixed or "historic" costs. Also, substitution of existing by new technology will usually be a gradual and not an instantaneous process. Despite this limitation, environmental biotechnology would very well satisfy the second criterion for the general diffusion of a new generic technology.

Social and political acceptability

Social and political acceptability are closely linked. They are key criteria for the success of major generic technologies. One of the main macro economic differences between telecommunication technologies and nuclear energy technology was that the first was socially and politically acceptable and hence economically successful whereas the second was not, in spite of earlier political support, which explains its ultimate failure in many countries. It has already been said that environmental biotechnology enjoys not only "political acceptability", but gets more – indirect – political support than any other branch of biotechnology through the legislative framework set up to protect the environment. This political acceptability is reinforced by a social acceptability which again seems to be greater for environmental biotechnology than for some other applications, at least in many countries.

128

The most explicit recent proof for differences in public opinion about various biotechnology applications came from Eurobarometer surveys financed by the Commission of the European Community in 1991 and 1993, and based on a large sample in all EC countries. In both years, the use of biotechnology research for environmental purposes is "agreed" or "strongly agreed" by 87 per cent of the public, a level higher than the support expressed for other applications (see also Figure 31 in Chapter 11).

Problems with public acceptance have been among the most important obstacles to a faster diffusion of biotechnology and particularly genetic engineering. Thus, the distinct advantage of environmental biotechnology in the eyes of the public is one of the most encouraging diffusion factors which, in the medium and long term, could have positive side-effects on the diffusion of biotechnology in general. The high degree of social acceptability could turn out to be no less decisive than the cost advantages of the new technology.

Environmental acceptability

The current techniques of environmental biotechnology, and particularly of bioremediation, do not yet require genetically modified organisms. Many scientists agree that for some years to come, there remains a wide scope for technological improvement through advances in the numerous sciences underpinning environmental biotechnology, even without genetic modification techniques. There are, however, scientists who emphasise that genetically modified organisms will soon become essential for the solution of more intractable environmental problems.

This raises the question of the acceptability of genetically modified organisms useful in environmental biotechnology. While this is not the place to review this issue, it is noteworthy that the national and international activities dealing with risk, regulatory and acceptance aspects of genetic modification seem to focus increasingly on bioremediation. While the current debate about possible releases of genetically modified organisms for environmental purposes is largely about safety, some aspects and expressions of this debate may have been complicated or exaggerated by matters other than safety concerns. These other matters include responsibility for control of a critically important and promising part of biotechnology. Little can be said about the likely outcome of this debate, except that the issue will probably not become critical for the diffusion of environmental biotechnology over the next three to four years.

Pervasiveness

More than any of the other four criteria, it is the pervasive effects of a new technology throughout the economic system which point to the emergence of a new "technological paradigm". In the past, biotechnology has sometimes been compared to information technology, whose influence can be felt in all economic sectors. However, it has also been emphasised that fundamental differences separate the two technologies, particularly the fact that biotechnology operates through living organisms or parts thereof, a fact which limits its field of activity to materials which can be biologically treated. This seems to exclude numerous industrial sectors, particularly those processing inorganic materials, from the direct influence of biotechnology.

This analysis does not take into account the "pervasiveness" of soil, water and air pollution, which could offer biotechnology an ubiquitous and global field of application, and one most likely to grow as long as the world's population grows. It is interesting to note in this context that environmental biotechnology is presently very much a domain of small and medium-sized, and not only large companies. Tens of thousands of companies across the OECD area are already involved, according to figures presented in Chapter 9. This reflects the "pervasive" or ubiquitous effects of pollution. In any event, environmental biotechnology is likely to become in the longer term the most pervasive branch of biotechnology, perhaps beyond health, agro-food or other applications (except for more futuristic forecasts of a fusion of biotechnology and information technology through biochips).

It has been argued that applications to health will be no less pervasive than to the environment, considering that virtually all humans are, or will be, in need of health care. Against this, one may reply that virtually all human activity is polluting in one way or another, most or all of the time, whereas the need for healthcare will not make itself felt for all humans simultaneously, and not for most or all of the time. Arguments for or against the greater pervasiveness of environmental biotechnology are likely to remain in the realm of speculation. Much will depend on the time perspectives chosen on relative cost advantages of health versus environmental biotechnologies (which seem to be more favourable to the latter), and on assumptions about long-term public and private expenditures for health and the environment.

At least from the perspective of industrial companies, (which may be a short- and medium-term view only), environmental biotechnology would seem to be more pervasive. Entry barriers seem to be low, with increasingly numerous and geographically close opportunities almost everywhere in the OECD area. This is why the number of companies is already much larger than that involved in health biotechnology.

The view of environmental biotechnology as the most pervasive application of biotechnology might be reinforced if one adopted the Japanese perspective which sees in it a key technology to address the global environmental problems of the 21st century: greenhouse effect, desertification, water shortage, resource depletion, etc. In that case, environmental biotechnology could certainly become a macro economic force with major impacts on productivity growth, investment and employment, beyond the size of current environmental markets (see Table 7).

What is then the answer to the question of whether the projected development of environmental biotechnology might lead to a modification of the conclusions drawn from earlier diffusion analyses? Certainly, in a long-term perspective of say, 20 years, environmental biotechnology could become the most dynamic and pervasive branch of biotechnology. This was not yet recognised a few years ago. The potential already exists, as the review of the five diffusion criteria shows.

However, this does not mean that the time it will take can be much shortened, or that the structural adjustments necessary to accommodate and encourage environmental biotechnology will be much easier or less than for other generic technologies. The educational and skill requirements for environmental biotechnology are a striking example of adjustment difficulties. One current problem which delays the introduction and diffusion of environmental biotechnology in industry is the traditional education of environmental engineers and managers which includes little or no biotechnology. On the other hand, biologists are not familiar with environmental engineering questions, which are rarely included in their curriculum. It is probably no coincidence that it is Japan, with its broad, future-oriented vision of environmental biotechnology, that has put education issues at the top of its biotechnology policy agenda.

Changes towards a better adaptation of the educational system to the requirements of environmental biotechnology are under way in other countries as well (Austria is an example). However, in spite of these adjustments, it must be said that major changes in curricula, training and skill profiles are generally a matter not of years, but, if not of decades, at least of one decade. This is also the period which is likely to pass before the more complex issues of fundamental science in environmental biotechnology are better understood and clarified.

Industry Strategies and Perspectives

Introduction

The following industry chapter of the overall study on ''Biotechnology for a Clean Environment'' compares the large number of available scientific and technological options described in Part I with existing industrial activities and perspectives. A review is presented and conclusions formulated on the industrial/economical perspectives of environmental biotechnology.

The study focuses on companies using biotechnology for environmental purposes in a broad sense, including detection, prevention, protection and remediation. A number of issues are raised which deal with economic and strategic aspects of biotechnology in the environmental sector. Each of these issues (see Appendix 1, Part II) raises in turn questions which this study addresses. These issues constitute the leading principle throughout this study: they were the subject of literature and data bank surveys as well as elaborate interviews and discussions with companies in this field.

As there is not one industry, even putting aside the various sizes, markets, market positions, R&D approaches and aims, there is, equally, no one industry view on environmental biotechnology. This is not a statistical analysis of data (which is hardly abundant for the environmental biotechnology sector) but rather a set of views, opinions and perspectives founded on a framework of data where possible, or in other words, a reflection of what industry is or is not doing in environmental biotechnology.

Chapter 7

Products and Services

Biotechnology will increasingly be applied within a broad range of industries and markets. Focusing on biotechnology as a means to attain sustainable development, a number of applications which differ in scope and purpose, can be identified. Table 9 gives an overview of the field of environmental biotechnology products and services.

The columns of Table 9 represent the three main stages of pollution control under headings which may be broken down further as follows:

- prevention of pollutants: process integrated technology, alternative processes, and clean technology;
- treatment of pollutants at the source: add-on technology, end-of-pipe technology, recycling technology, abatement/control technology;
- treatment of diffused pollutants: remediation, restoration, valorisation.

The rows refer to the three environmental compartments: air, water and soil. These are the compartments through which pollutants are transported or accumulated into the environment. The fourth and last compartment in the horizontal row is solid waste. This category includes dredging and harbour sludge, industrial waste (including chemical and power station waste), household waste, sewage sludge and surplus manure. This compartment-based structure reflects the organisation of industry and legislation, as well as differences in technologies and approaches to treat pollution.

In theory, environmental products and services can be found in each of the cells of the matrix of Table 9. In reality, if we look at the market for environmental biotechnology products only four cells (shown by the "+") contain one or more substantial products/services at present. Most of the existing markets for environmental biotechnology products can be located in the second and third row. In general, however, industry expects environmental biotechnology to grow in all cells.

In addition to the two dimensions in the matrix of Table 9, two other dimensions can be added: place/position (local versus global) and time (now versus future) to complete the overview of the environmental problem (see Table 10). Whereas the concept of sustainable development, including the role of environmental biotechnology, is directed at global solutions which can only be realised in the longer term, most industrial activities are positioned in the local and short term segment in Table 10. This is where continuation of their economic activities is located. Most individual companies concentrate their efforts on making their processes less polluting or on expansion if they look away from the "here and now". They expect international governmental

Table 9. **Overview of environmental biotechnology activities**

Compartment	Prevention of pollutants		Treatment of pollutants at the source		Treatment of diffused pollutants		Detection and monitoring	
Air/off-gas	–	(+)	+	(+)	–	(–)	–	(+)
Water	+/–	(+)	+	(+)	+/–	(+)	+/–	(+)
Soil	+/–	(+)	–	(+/–)	+	(+)	–	(+)
Solid waste	+/–	(+)	+/–	(+)	+	(+)	–	(+)

Note: The most relevant combinations for industry are indicated by "+", less relevant combinations by "–". A possible future situation is given between ( ).
Source: P. Hesselink, TNO, Delft, Netherlands.

Table 10. **Scale and time frame of environmental biotechnology activities**

Scale	Short term	Middle and longer term
Local $\longrightarrow$	Industrial activities $\longrightarrow$	Cleaner industrial activities
	$\downarrow$ $\searrow$	
Global	Industrial expansion	Sustainable development

Arrows indicate the focus of development.
Source: P. Hesselink, TNO, Delft, Netherlands.

actions to pursue more sustainable development world-wide. However, the focus of countries differs with respect to Table 10.

In order to move from local and short-term measures of companies to (more) sustainable development in the future, several activities should be carried out by governments. Industries' opinions in this respect are given in Table 11.

Clean production and pollution prevention technology can be roughly divided into two main categories: process integrated and add-on technology. The former is rather small in comparison to other environmental (bio)technology and is therefore discussed first.

In the other parts of this section a presentation of the large majority of environmental biotechnology products and services is given arranged according to the compartments air, water, soil, and solid waste. For these four segments the impact and the competitive advantages and shortcomings of biotechnology compared to other environmental technologies are discussed.

The role of biotechnology in detection and monitoring of pollution, with respect to industrial activities or application in the market place is very small at present. Only a few products are commercially exploited. However, much R&D is currently undertaken, and some products are at prototype stage.

Table 11. **Government roles in facilitating the development of environmental biotechnology according to industry**

Scale	Short term	Middle and longer term
Local	* Consistent legislation	* R&D support/stimulation
	* Consistent enforcement of legislation for state owned, local and multinational companies	* Facilitate innovative technology demonstration (to enhance reliability and predictability)
	* Standards and legislation which do not change over time to allow adequate return on investments	* Stimulate education (universities)
		* Dissemination of results
	* Tax deduction for environmental investments	* Motivation of public
Global	* Harmonised legislation (competiton factor)	* International co-operation and planning in:
	* Set priorities/aims	– R&D
	* Action plans	– Actions
	* Seek and encourage the worldwide diffusion of technologies through which future economic growth can be achieved with ever increasing efficiency in the use of materials and energy, and progressively diminishing impacts upon the environment	

Source: P. Hesselik and C. Enzing, TNO, Delft, Netherlands.

1. Process-integrated environmental biotechnology

Prevention of pollution by process-integrated biotechnology is an important and growing field but is hardly visible in the sense of concrete products or services; rather it is an integrated part of process development and design. Although it is easy to indicate the size of the market for particular types of filters or scrubbers, it is not possible to discern a market for "clean technology" (the first vertical category in Table 9). This has implications for economic analysis and comparison, because as the means of controlling pollution are merely incorporated into the production process, so are the costs. An indication of the amount of process-integrated investments as a share of total environmental investments is given in Tables 12 and 13. An overview of some industrial projects with respect to their environmental effects in France is given in Table 14.

Table 12. **Share of integrated technology including biotechnology
in pollution control investment**

Country	Share of integrated biotechnology (%)
Belgium	20[1]
Germany	18[2]
France	13[3]

1. Interenvironment Wallonie (Flanders excluded) 1988.
2. IFO Institute (private investments only).
3. Ministry of the Environment, 1987.
Source: EC, 1992.

Most of these projects deal only with end-of-pipe type of treatment, and process integrated measures (fundamental changes in the production process) represent only a small proportion. Biotechnology constitutes an even smaller part at present. Although an increase in the share of process integrated (bio-)technology for clean production is foreseen by industry, this is unlikely to take place in the short or medium (5-10 years) term.

In general 'clean technology' is in an early stage of development. Very recently, industries (mainly the petrochemical and pharmaceutical multinationals) have started R&D projects aiming at developing clean technologies to meet anticipated future legislation. Although the advantages of biotechnology-based production and processing are generally accepted, these processes are currently only used in the food industry which is by tradition a biology-based industry, and in some parts of the pharmaceutical industry (pharmaceuticals and fine-chemicals).

There are a number of practical and cultural reasons why the diffusion of biotechnology in process integrated technologies moves only slowly in, for example, the chemical sector. The period for depreciating a chemical production plant is rather long and not an incentive for substantial changes. Once a large plant has started to generate products and money, it should do so and not be disturbed by drastic changes in the process. Furthermore, the costs for approval of introducing/replacing (new) parts in the production process are very high. In most European countries as well as in the United States the procedure for the approval of a proposed alteration in the production process of pharmaceuticals lasts almost two years and is expensive even if it is only an improvement from an environmental viewpoint. This is often given as a reason for not changing an approved process.

Table 13. **Relative size of investment in clean technology for air and water**

	United States – 1985 (million US$ and %)	France – 1986 (million FF and %)
Total expenditure *of which:*	2 310	4 200*
Add on	1 801 (78%)	3 630 (87%)
Process-integrated	509 (22%)	570 (13%)

* Including solid waste.
Source: EC, 1992.

Table 14. **Environmental projects in France**

Pollution reduction	
Water	95% of cases
Air	5%
Waste water	8%
Savings on consumption	
Water	70%
Raw materials	60%
Energy	13%
Upgrading of waste	26%
Reducation of accidental risks	25%
Improvement of working conditions	30%

Source: Clean Technology Task Force – France.

Nevertheless, the companies in question are mostly multinationals and already have biotechnological expertise in house. Most of the companies interviewed spend an increasing amount of their R&D effort on the development of bioprocesses, but only for future and not for existing processes. For small- and medium-sized companies, changes in production processes have a large effect. These companies will generally hesitate to introduce new technologies unless they have been demonstrated as cost-effective and reliable. Financial and scientific support by third parties or government will therefore be important. Spin-offs of such support will only be introduced in industrial practice in the longer term.

In general, industry expects process-integrated measures to become relatively more important in the long term whereas clean-up and treatment/abatement technology is indispensable both for the short, medium and long term.

2. Air/off-gas

Air/off-gas treatment includes the production of equipment, reagents and services needed to abate gaseous pollutants or reduce their level in flue gases (the main part of the industry), in ventilation or process air (for example, solvent-contaminated air from surface coating and stripping processes) and in subsoil air from site clean-up (contaminated air is extracted from the soil and cleaned).

The main contaminants which have to be removed from industrial units and power stations are SO_2, NO_x, particulate materials, CO, CO_2, odours, H_2S, partially or non-oxidised hydrocarbons including aromatics, halogenated hydrocarbons and all volatile organic compounds (VOCs).

The main air treatment processes used in industry at present include desulphurisation, denitrification and mechanical dust removal systems. A distinction is made between primary processes that reduce the contamination during the incineration process and secondary ones that reduce contamination and the exhaust level of the flue gas.

Impact of biotechnology

The biotechnological processes used in air/off-gas treatment are primarily biofiltration, followed by biological sorption processes such as bioscrubbing and biotrickling filtration. They constitute the most important specialised processes in this subsector.

Biofiltration and bioscrubbing are widely applied (Netherlands and Germany) in industries and processes in which odorous gases are generated. They were successfully introduced in the food industry, breweries, some chemical processes, waste water treatment units, agriculture, etc. This led to rapidly increased sales of biofilters (Figure 20). Furthermore, biofilters have been used to remove easily biodegradable compounds emitted by oil cracking or off-gases from the petrochemical industry and the feed and food industry. Many of these off-gases were previously treated by physical or chemical air treatment systems. Reasons for treatment are either local complaints or new legislation.

Nowadays, many pollutants with a concentration lower than 1 000 mg/m³ can be treated biotechnologically and an extension of the application area to other types of pollutants in higher concentration ranges is expected (Figure 21).

Figure 20. **Capacity of biofiltration installed per year by two major Dutch companies, 1984-90**

Gas flow (thousands m³ h⁻¹)

Gas flow (thousands m³ h⁻¹)

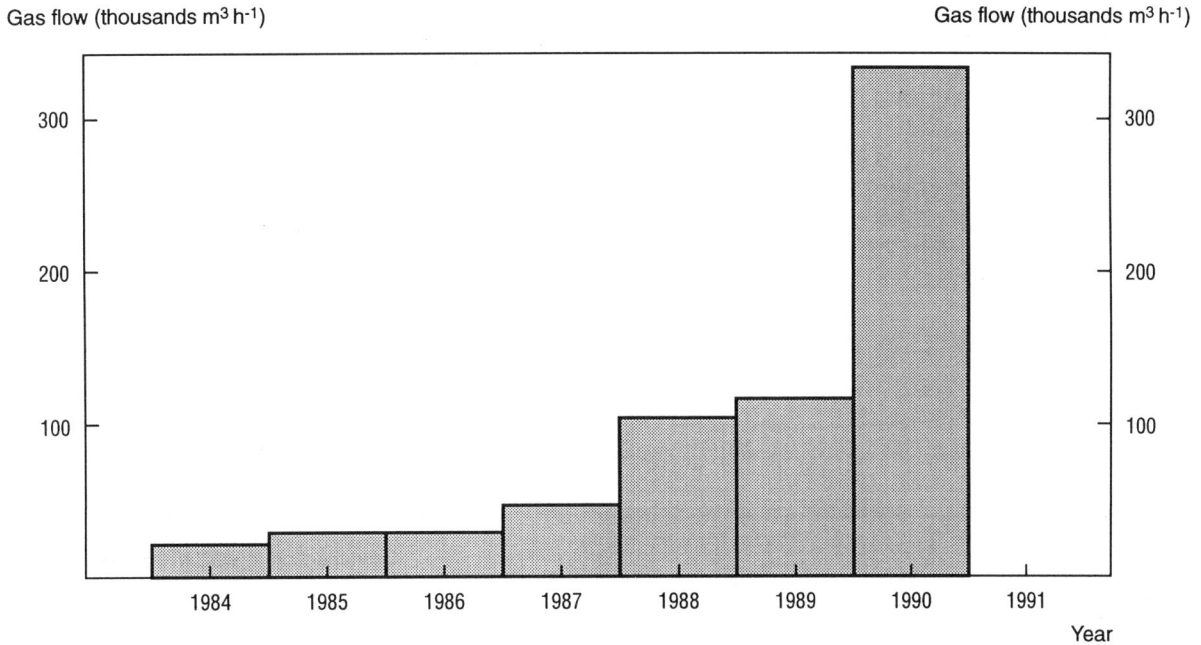

Source: Van Groenestijn and Hesselink, 1994.

Figure 21. **Economically feasible application area of various techniques for air pollution control**

Gas flow (m³ h⁻¹)

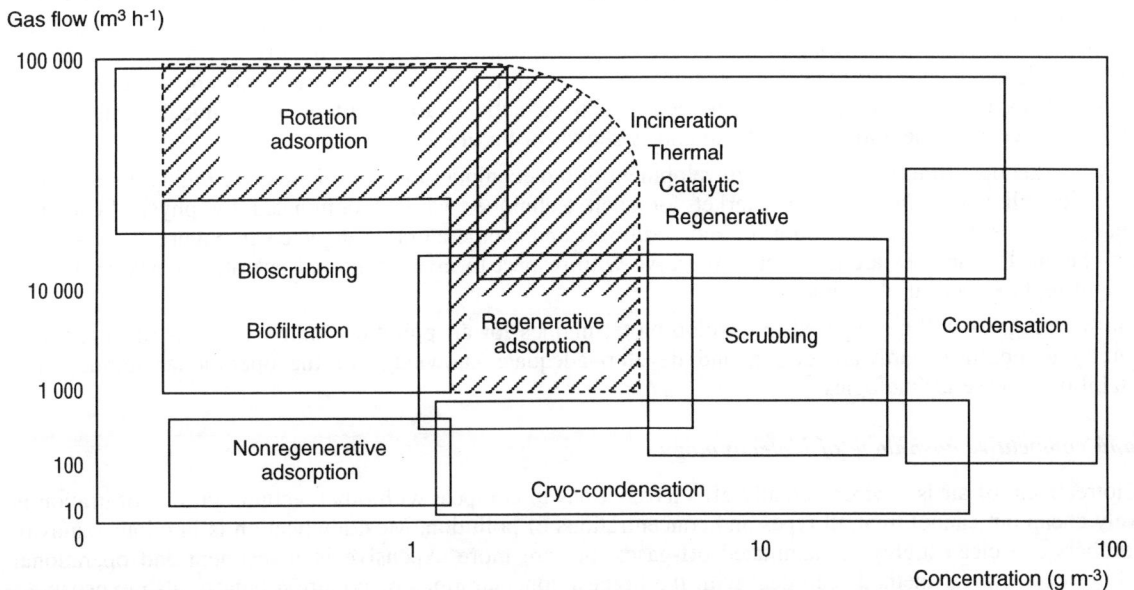

Concentration (g m⁻³)

Note: Biotechniques like biofiltration and bioscrubbing are expected to extend their fields of application into the shaded area in the near future.
Source: Van Groenestijn and Hesselink, 1994.

Figure 22. **Comparison of various biotechniques for air pollution control**

Operational costs (Dfl g^{-1} VOC)   Operational costs (Dfl g^{-1} VOC)

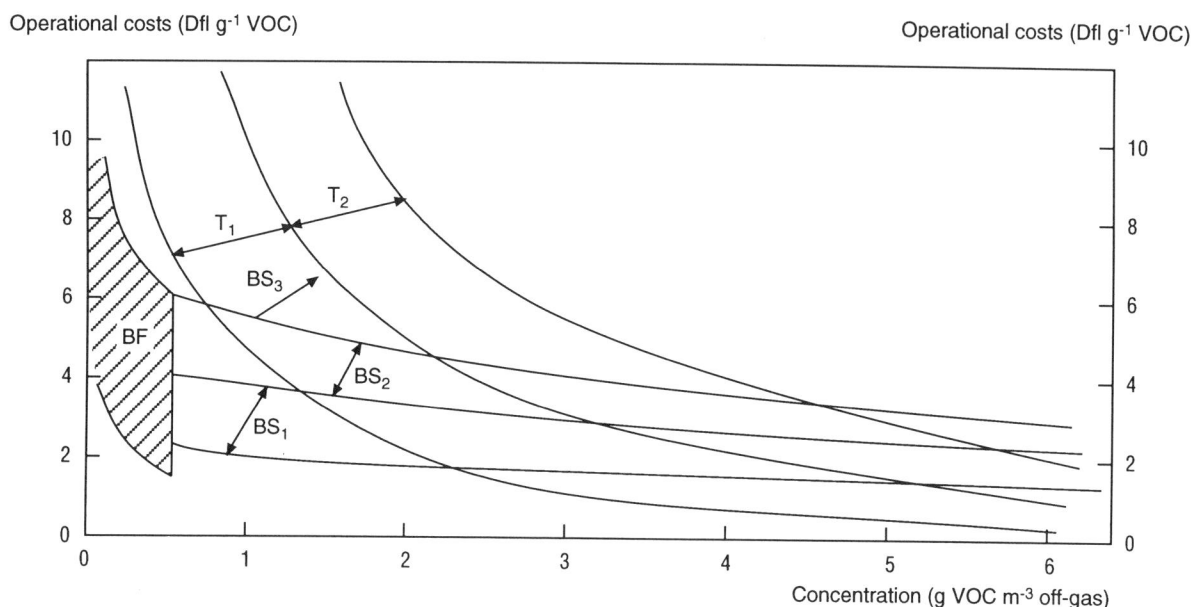

Concentration (g VOC m^{-3} off-gas)

Notes: BF = Biofilter.
      BS = Bioscrubber of first to third generation.
      T = Thermal treatment (incineration) of first or second generation.
Source: Van Groenestijn and Hesselink, 1994.

A second generation of biofilters and also bioscrubbers and biotrickling filters for industrial air/off-gas treatment should shortly reach the market. These processes employ specialised micro-organisms and/or combinations of biological with physical and chemical techniques such as membrane technology. Such filters can treat higher concentrations of pollutants and toxic pollutants *e.g.* denitrification for the removal of ammonia from manure, photocatalytic pretreatment, or styrene removal from off-gases using moulds. As a consequence of technological development, the costs for cleaning per m^3 of air/off-gas may increase, but are not expected to exceed the costs of using competing technologies such as active carbon filters or chemical scrubbers. A comparison between incineration and biotreatment is shown in Figure 22.

New market opportunities include the development of biotechniques capable of cleaning highly concentrated and/or toxic pollutants. The potential market for such techniques is now dominated by physical/chemical techniques such as active carbon filtration and scrubbers and incinerators as depicted in Figure 21. As these physico-chemical techniques are relatively expensive with respect to investment and operation costs, biotechniques might replace them in due time.

Further progress will be required to develop biotechniques for air pollution control to industrial standards, particularly to optimise hardware design and develop adequate knowledge of the operational stability and predictability of these technologies.

Cost and competitive advantage of biotechnology

Biotreatment of air is environmentally effective but has to compete with other technologies. Biofiltration is relatively cheap but cannot treat all types and concentrations of pollution. Most application is in odour removal. Bioscrubbers can clean highly contaminated off-gases, but are more expensive in investment and operational costs. Physico-chemical methods can deal with the highest concentrations of pollution but are also expensive.

Biotechniques require little in the way of energy or chemicals and produce no or only small amounts of secondary wastes. An overview of indicative costs, energy consumption and materials balance is given in Table 15. It is clear that biotechniques are competitive with other techniques with respect to all these issues.

Table 15. **Indicative costs, energy consumption and materials balance of some air treatment techniques**

Technique	Raw materials	Waste materials and emissions	Energy requirements (kWh/m³)	Costs (Dfl/1 000 m³)
Biofiltration	Compost, water	Compost, CO_2, water and organic degradation products	≤ 1	0.50-5.00
Bioscrubbing	Water	Biomass, CO_2, water and organic degradation products	1.6-3.0	3.00-6.00
Scrubbing (chemical)	Water or solvent + chemicals (chemicals for regeneration)	Polluted water or solvent, concentrated contaminants, polluted regeneration solvent	0.1-10 5-10 (for) regeneration	1.00->20.00
Adsorption (by active carbon)	Active carbon (regeneration solvent)	Active carbon with contaminants, polluted regeneration solvent	100-200 kWh/ton VOC 2-4 ton of steam/ton VOC for regeneration	1.00-10.00
Incineration	(fuel)	Exhaust gases	$1\text{-}3 + 0\text{-}2\bullet10^6$ kJ/1 000 m³	2.50-25.00
Catalytic incineration	Catalyst (fuel)	Exhaust gases and split/poisoned catalyst	$1\text{-}2 + 0\text{-}10^6$ kJ/1 000 m³	2.5-20.00

* Indicative costs for Dutch situation.
Source: Hesselink and Stoop, 1993.

Advantages of biotechnological treatment of air/off-gas include:

– pollutants are totally converted into harmless substances without accumulation of toxic residues or side products;
– lower operating costs (operating costs for biological gas treatment typically amount to 40 per cent and 20 per cent of those associated with thermal and chemical processes respectively);
– favourable energy balance;
– favourable balance of nutrients (chemicals).

Disadvantages include:

– cannot be used for all pollutants;
– largely empirical processes based on trial and error/black boxes (most biofilters are now based on undefined complex microbial populations);
– largely unknown to "classical" off-gas control engineers.

3. Water

The waste water treatment industry provides reactants, equipment, installations and services applied to drinking water, domestic sewage, process water, industrial waste water and groundwater. Water treatment is a heterogenous technology requiring such diverse measures such as the construction of facilities, microelectronics for control, (chemical) engineering, and modern biotechnology. Depending on the type of water to be processed, there are many possible treatment methods which are combinations of biological, physical and chemical techniques. Biological treatment techniques are both aerobic and anaerobic, and each offers different advantages and fields of applications. Combinations of aerobic and anerobic processes or combinations of biological and physico-chemical methods are increasingly important.

Waste treatment is by far the most developed sector of air, water, soil and solid waste treatment (see also Chapter 8). In many OECD countries, over 50 per cent of the waste water was already treated at the end on the 1980s and this amount is still growing (Table 16). This holds both for domestic and industrial waste water.

Industrial waste water treatment is carried out in municipal as well as in specific industries treatment facilities. For example, about 49 per cent of the industrial waste water in the Netherlands is treated in municipal facilities and 51 per cent is treated in industry-owned installations.

A general trend in discharge rules and regulations is towards increasingly tighter levels, imposing the necessity of more complete water treatment to remove pollutants. It should be noted that even low toxicity pollutants (*e.g.* phosphorus, nitrates, and/or salt) must be more and more severely reduced. This imposes

Table 16. **Primary to tertiary waste water treatment in OECD Member countries**

(in per cent of total population served)

	1975	Total served 1980	1985	1987-1989
Belgium	6	23 (1979)	36	42
Denmark	71 (1977)	–	91 (1986)	98
Germany	75	82 (1979)	87 (1983)	90 (1987)
Greece	–	1	–	–
Spain	14	18	29	48
France	31	43	49 (1984)	51
Ireland	–	11	–	–
Italy	22	30	–	60
Luxembourg	–	81	83	91
Netherlands	45	72	85	91
Portugal [1]	6	9	9	11
United Kingdom [2]	–	82	83	84
USA [3]	67 (1976)	70	74 (1984)	–
Japan	23	30	36 (1984)	39

1. EC estimates to 1985.
2. England and Wales only.
3. 1980 data include 1 per cent and 2 per cent of non-discharge treatment. 1980 and 1985 data were determined by using different methods than previous data and therefore may not be comparable.
Definitions: Primary treatment: removal of gross solids. Secondary: removal or organic material or bacteria under aerobic conditions. Tertiary: removal of suspended solids following secondary treatment.
Source: EC, 1992.

technological challenges on the efficiency of the overall process. Furthermore, process economics of water usage and treatment are also increasingly important and are improved by recycling after treatment or by recovery of ''by-products'' such as biogas.

As a consequence, water usage has generally become more efficient as the data in Figure 23 shows.

Figure 23. **Industrial water consumption in Japan, 1965-85**

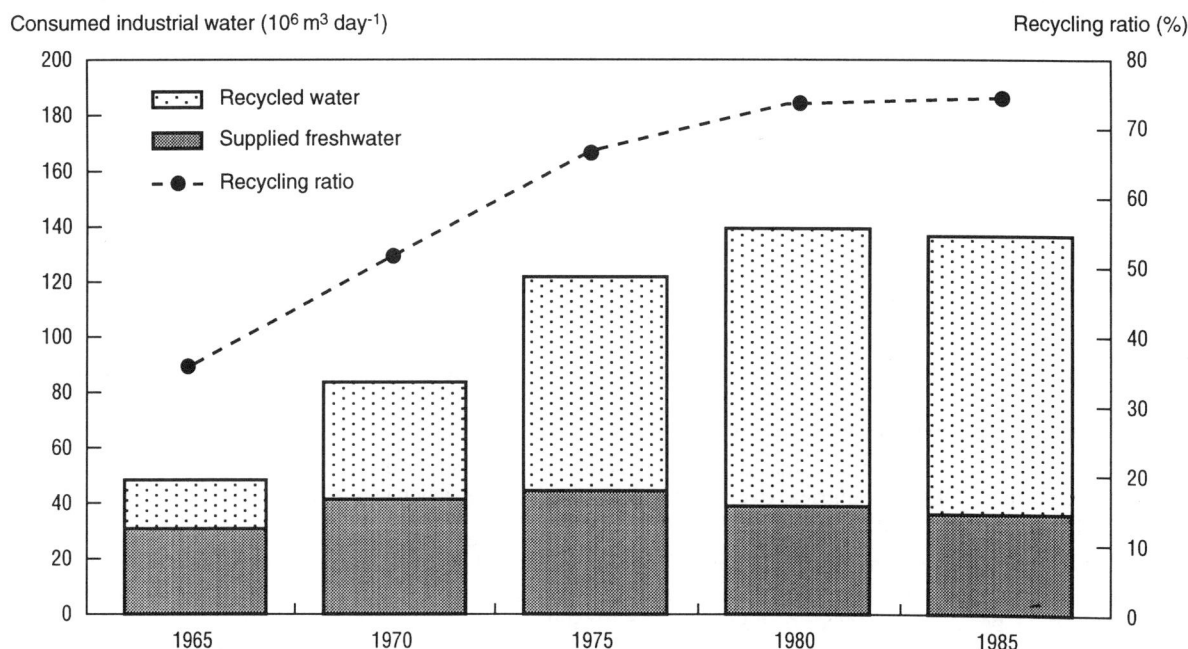

Consumed industrial water ($10^6 \, m^3 \, day^{-1}$)

Recycling ratio (%)

- Recycled water
- Supplied freshwater
- Recycling ratio

Source: Water Science and Technology, Vol. 23, 1991.

Impact of biotechnology

The share of biotechnology in this market is rather high, because micro-organisms are extremely suitable for the treatment of waste water containing the more common organic pollutants. This is true both for domestic for industrial waste water, especially for those industries which process materials of agricultural origin. The value of biological treatment was already realised more than a century ago when innovative R&D was first applied in this sector. As a result, both aerobic and anaerobic waste water treatment systems have been developed for use in municipal and industrial applications.

Until the beginning of the 1980s, mainly aerobic systems were used, mostly active sludge installations or oxidation beds. During the last ten years, anaerobic systems have become more and more used for waste water treatment, often in combination with aerobic processes. In the purification of waste water, biotechnological methods are already widely used for the removal of nitrate, phosphate, heavy metal ions, (chlorinated) organic compounds, and toxic substances. New techniques are being developed for the treatment of industrial waste water with special contaminants, which are particularly relevant in the following industries: industries using catalysts, petro-chemicals, chlorine, textile, food/feed and leather production, cellulose and starch processing, electroplating, mining, surface degreasing and coating, and and printing.

In the near future a number of industrial developments based on biotechnology are likely to become implemented in the field of water treatment:

- increase of operational stability and reliability by improvements in monitoring, control and bioprocess technology;
- removal of nitrogen, phosphorus and sulpher compounds and micropollutants;
- extension of bioprocessing to various industrial processes (*e.g.* the petrochemical and chemical industry) and more special processes (*e.g.* coal tar distillation where the use of peroxidases/oxygenated water systems is actively being investigated);
- application of specialised, highly active strains (to increase overall productivity) for the treatment of specific pollutants through bioprocessing.

Japan is also exploring additional issues including immobilisation by entrapment, symbiotic systems, biological membranes, and biodegradable flocculants.

These developments in biology will be primarily based on the selection and engineering of improved microbial consortia capable of withstanding higher pollutant concentrations and with a higher biodegrading capacity. In the longer term, the engineering of enzymes with an adequate level of stability for the industrial degradation of specific pollutants is foreseen. Technological R&D will focus on reliability and predictability of processes. Both intensive (compact, high space/time yield) as well as extensive (cheap, long treatment times) processes will be developed and applied.

Cost and competitive advantages of biotechnology

Bioprocessing plants have until now been used mainly for waste water treatment from industrial processes with a high carbohydrate content such as cellulose or starch: an extension of the application area of biotechnological waste water treatment to other industrial effluents is currently taking place.

For instance, biosorption may replace physical/chemical methods like precipitation, complexation, adsorption or ion exchange in scavenging heavy metals ions including radioactive compounds from water. The advantage of such a technique is that extremely dilute solutions can be purified at low costs. In addition, the recycling of As, Cd, Cr, Pb and Hg may be possible because accumulation to high concentration is possible (up to 12 per cent microbial dry weight).

The application of biotechnology in waste water treatment faces a number of obstacles for further development:

- The regulatory framework has an influence on any environmental market, including the market for environmental biotechnology. This is particularly true for the removal of specific pollutants. Depending on each country's geographic and climatic conditions, different measures can be introduced, representing different stimuli for bioremediation. In this respect the fundamental role played by some north-west European governments to lower the maximum allowable nitrogen concentration in waste water within a relatively short period should be underlined.
- The development and application of new and improved microbial strains or consortia could be difficult as water treatment units are typically open systems, where it could prove to be impossible to maintain an engineered strain in equilibrium.
- Operational stability and predictability.

Table 17. Indicative costs, energy consumption and materials balance of some water treatment techniques

Technique	Raw materials	Waste materials and emissions	Energy requirements (kWh/m³)	Costs* (Dfl/m³)
Biological water treatment	(Oxygen)	Biomass, biogas, CO_2 and organic degradation products	0.1-0.3	0.10-1
Sedimentation	Chemicals	Precipitated sludge (polluted)	0.1-0.3	0.05-30
Flotation	Chemicals and/or air	Flotation sludge, odours	0.1-0.4	0.10-2
Adsorption	Adsorbent (solvent for regeneration)	Polluted adsorbent (polluted regeneration solvent)	0.1 (100-200 kWh/ ton VOC for regeneration)	1-10
Chemical oxidation	H_2O_2, O_3, UV	CO_2, N_2, degeration products	2-25 kWh/kg (O_3, H_2O)	0.50->5
Micro/ultra-filtration	–	Concentrated wastewater	0.5-20	<1->20

*   Indicative costs for Dutch situation.
Source: Hesselink and Stoop, 1993.

- The application of recombinant microbial strains could raise questions of safety analysis and management, and of public concern.
- Most waste water systems are "tailor-made" systems. Transfer of the same system from one application to another is difficult.

These obstacles may delay the further diffusion of biotechnology in this industry and emphasises the need for more R&D in bioprocess technology and microbial ecology. Nevertheless, biological treatment has become dominant over other technologies in this sector for reasons of cost effectiveness and relative ease of operation combined with a long use as proven technology (Table 17).

Advantages of biological treatment of waste water:

- both very high and very low concentrations of (organic) pollutants can be handled;
- in anaerobic systems biogas is produced;
- low costs per m³ of treated water;
- use of biosorption or bioprecipitation techniques will enable concentration and recycling possibilities of *e.g.* heavy metals, for example.

Disadvantages include:

- sludge production;
- operational stability, predictability and controllability of industrial waste water plants is limited;
- not all non-organic pollutants can currently be removed;
- translation of biological processes into biotechnological concepts that can be used on an industrial scale is making slow progress, and efficient, easily controllable and multi-purpose reactors are still needed.

4.   Soil

The activities in this segment mainly deal with site clean-up and reclamation including, for example, the remediation of industrial sites, spills and abandoned landfills. Contrary to air and water treatment, services such as soil cleaning rather than systems or installations are offered. Services comprise *in situ* treatment (without removing soil) and *ex situ* treatment in (bio-)reactors and/or landfarms.

Impact of biotechnology

Microbial soil treatment is a technique which has already passed the first experimental stage. In essence it is the capability of bacteria and fungi to use the contaminants as a food- and/or energy source. Just by optimising the physiological conditions in the soil (either *in situ* or *ex situ*), the capability of micro-organisms to demolish pollutants such as cyanides, polycyclic aromatic hydrocarbons (PAH) and other organic molecules will increase.

Table 18. **Costs of three *ex situ* soil treatment techniques**

(Germany)

Method	Cost (DM/tonne soil)
Thermal (incineration)	80-450
Chemical/physical	20-200
Microbiological	50-150

Sources: Raucon, 1989, and Reiss, 1990.

The efficiency of microbial soil treatment techniques depends on the following factors:

– the biodegradability of the pollutants;

– the availability of the pollutants to the micro-organisms (bioavailability);

– the conditions (*e.g.* oxygen, humidity, temperature), which facilitate the biological treatment process;

– the presence of a population of micro-organisms with appropriate biodegrading properties.

Experience in the United States shows that 65-85 per cent savings occur when biological methods are used instead of physical/chemical procedures. Whereas incineration usually costs $250-500 per tonne of soil, biological methods can cost as little as $40-70 per tonne (see also Table 18). With improved bioprocesses, costs will be further reduced in the future. This particularly applies to projects undertaken to clean oil-contaminated soils (150 000 cubic feet in up to 10 weeks).

Compared to other technologies (*e.g.*incineration, landfill), the actual application of biotechnology is still limited, but it is rapidly growing. Biological treatment offers particularly elegant solutions for site reclamation, especially for the treatment of complex organic contaminants and moderately contaminated sites whenever disruption of the industrial complex or real estate and its infrastructure (piping, etc.) is expensive or impossible. The main impact of biotechnology in the near future is expected to be in cleaning up old industrial sites (former refineries and gas works, airports, petrol filling stations and abandoned landfills).

Figure 24. **Microbiological *ex situ* treatment of 300 tonnes of soil of a refinery polluted with hydrocarbons (HC)**

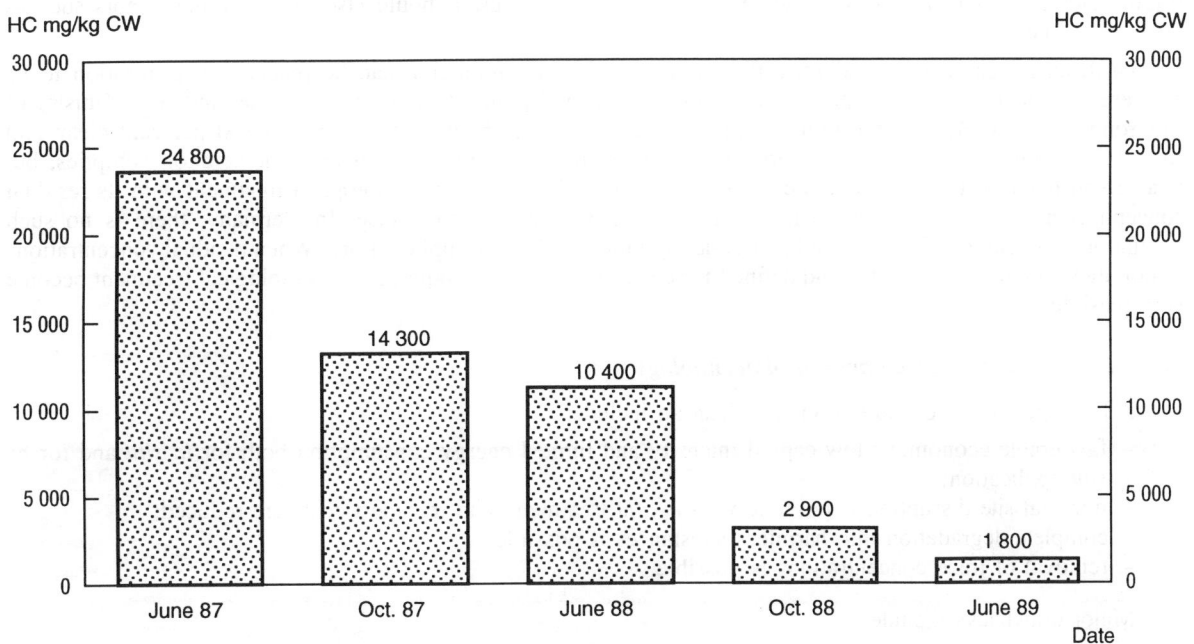

HC mg/kg CW

HC mg/kg CW

Date

Source: Umweltschutz Nord, 1991.

Table 19. **Indicative costs, energy consumption and materials balance of some soil remediation techniques**

Technique	Raw materials	Waste materials and emissions	Energy consumption (kWh/tonne)	Costs* (Dfl/tonne)
In situ				
Biorestoration	Water, oxygen, compost	CO_2, biomass + organic degradation products, water	20	70-150
Chemical extraction	Adsorbent, water (regeneration solvent)	Polluted water, polluted adsorbent (and regeneration solvent)	20	125-150
Electro-reclamation	Rinsing solution (electricity)	Polluted rinsing solution, concentrated contaminants	100-500	150-300
Steam stripping	Water (fuel)	Polluted water and air	85-200	250-300
On/off site				
Land-farming	Fertilizers	CO_2, H_2O	10-30	50-140
Chemical extraction	Adsorbent, regeneration solvent	Polluted adsorbent and regeneration solvent, sludge	20-30	120-240
Thermal treatment	(fuel)	Exhaust gases, "dead soil"	250-700	100-300

* Indicative costs for Dutch situation.
Source: Hesselink and Stoop, 1993.

Industrial R&D in this segment is now mainly focused on the treatment of complex mixtures of contaminants, the increase of reaction rates, bioavailability, screening and selection of micro-organisms able to degrade compounds such as chlorinated hydrocarbons, dioxins, PCBs, pesticides, etc. The developments in this field are based mainly on strain selection, strain engineering and on-line monitoring using detection techniques such as biosensors. In the near future, integrated hydraulic isolation and *in situ* bioremediation combining biological with physico-chemical methods such as percolation processes will become available.

Costs, reaction rates and results (safety, completeness and control of the proces) will determine the final choice for biological or other techniques. When only costs are considered one should chose microbial techniques for *ex situ* soil treatment. Table 18 provides a German cost comparison for some *ex situ* techniques.

Considering process rates alone, incineration is favoured since biological treatment takes considerable time (Figure 24). However, decisions on which technique should be used should also include other factors such as those mentioned in Table 19.

With each treatment procedure, different levels of decontamination can be reached. The question to be answered is which level is acceptable from an environmental point of view. In the Netherlands the Ministry of Environment (VROM) uses the rather stringent A-norm for the final results: close to 100 per cent clean and suitable for multifunctional use after treatment. Given this performance level only incineration complies, but treated soil has lost its structure and cannot be used as soil anymore. For biological treatment with its residual concentrations of contaminants, such A-norms are generally hard to realise. In Germany, there is no such regulation concerning final levels and consequently biotechnology is applied more. When residual concentrations of pollutants are legally accepted and defined for clean soil, other techniques such as biological treatment become very feasible.

Costs and competitive advantages of biotechnology

Advantages of biotechnology in site clean-up include:

– favourable economics: low capital intensity and limited energy consumption both for *in situ* and for *ex situ* application;
– minimal site disruption is possible with *in situ* treatment, and transport costs are reduced;
– complete degradation into harmless substances is possible;
– removal of trace concentrations is possible.

Major drawbacks include:

– complex mixtures of contaminants are difficult to treat;
– rate of reactions is still limited;

144

- translation of laboratory conditions to technology at large practical scale;
- selection of the suitable microbial strains is largely based on trial and error;
- residual concentrations and bioavailability problems.

5. Solid waste

The main technologies for handling solid waste are landfilling, recycling, volume reduction, biological treatment (including composting), chemical treatment, incineration (with or without energy recovery) and disposal in the sea. Generally, solid waste contains a mixture of contaminants, present in several matrices such as landfills or industrial waste dump sites. Remediation of such a variety of problems requires an appropriate strategy for each hazardous or industrial waste situation for which the whole range of biological and non-biological technologies available must be used.

Impact of biotechnology

The current industrial application of biotechnology is largely limited to composting of wastes containing vegetable organic fractions. Future applications are to be expected in the following fields:
- Detoxification of solid waste (*e.g.* selective removal of heavy metal ions) allowing the selective concentration of pollutants present only in trace amounts. Developments in this field could be stimulated by advances in genetic engineering and molecular modelling allowing the creation of selective and high affinity ligands for these pollutants.
- Digestion of wastes with an industrial organic (*e.g.* oil, solvent) content. Until now these wastes are only slowly degraded during the composting process, leading to contamination problems of groundwater.
- Transformation of waste into biogas (*cf.* waste water treatment) and overall process intensification, allowing a more rapid waste turnover.
- Development of biodegradable plastics is being explored in all major OECD Member countries as a way to decrease the amounts of solid waste.

Most of these developments will derive both from improvements in the existing biological systems and from technological solutions. These improvements include advances in microbial strain selection and strain engineering as well as improved hardware design, particularly to reduce residence times and overall energy requirements. The general technological trend here is towards sophisticated technical solutions such as biological or thermal waste treatment combined with energy recovery, as other solutions are increasingly becoming difficult (*e.g.* landfill disposal). Alternative solutions such as recycling are increasingly applied. Recycling is particularly favoured as a result of increasing awareness of resource conservation, pressure on landfill disposal, and saturation of existing landfill sites. This method is used already for waste types such as paper, glass, metals and plastics in various OECD Member countries.

Competitive advantage of biotechnology

Advantages of biotechnological treatment of solid wastes:
- selective concentration and removal of components is possible;
- operating (and investment) costs are moderate;
- overall waste volume is significantly reduced;
- options for adding value, reuse or conversion into products with a positive value (*e.g.* fuel).

Disadvantages of biotechnological treatment of solid wastes:
- reaction rates are still slow;
- some wastes have a limited susceptibility to biological degradation;
- mixed microbial populations are needed to handle complex compositions, but this should not always be a disadvantage.

6. Conclusion

The overall impact of biotechnology on the environmental sector varies from substantial in one segment (water) to limited in other segments. However, the potential of biotechnological techniques is recognised. Industry expects an increasing role of biotechnology in environmental management, and biotechnological techniques are recognised to have a considerable number of advantages compared with other environmental techniques.

Biotechnology is unlikely to be the sole solution to environmental problems but will rather be a complementary tool to be integrated within a broader set of technologies in order to develop turnkey systems, focused at solving environmental problems where what counts is not the technology used but the possibility of a solution.

However, key to further development and application of biotechnology in treating environmental problems will be the integration of many disciplines from basic (micro-) biology to bioprocess engineering, a broadening of the range of pollutants which can be biodegraded and, most important, an improvement of the operational stability, reliability and predictability of environmental biotechnology processes. Only if environmental biotechnologies reach this level of cost-effective proven technologies, will they be applied by industry on a large scale.

Chapter 8

Market for Environmental Biotechnology Products:
Supply, Demand and Competition

1. Introduction

This chapter considers the market forces in the various segments of the European market for environmental biotechnological products. Before going into more detail, the general developments in the environmental market, of which environmental biotechnology activities are an integral part, are reviewed. Next a number of aspects of the bio-environmental market are discussed: market size and dynamics, the suppliers to the market, competition and competition factors.

On the supply side are the companies offering environmental biotechnology products and services. On the demand side, the environmental biotechnology market comprises a wide variety of industries, each with specific pollution problems and hence with requirements for specific goods and services. It is the mismatch between highly fragmented supply side and the specific, generally legislation-based needs of the demand side which is providing the dynamics for much of the activity being observed within the (bio-)environmental market in Europe at the present time.

2. Market size and forces

In a previous OECD study, the environment market was described as "including firms which produce pollution abatement equipment and a range of goods and services for environmental protection and management". This broad definition of a wide variety of activities does not allow clear and detailed statistical classification and leads to various interpretations of the data. However, market trends of the environmental industry can be distinguished with some reliability. In general, about 75 per cent of the environmental market consists of equipment production and its related services for mainly add-on or clean-up purposes. The remaining 25 per cent consists of general services related to technical engineering, environmental consulting and management services, which comprise both add-on/clean-up technologies as well as process integrated/clean technologies.

Industry Structure

In the abovementioned OECD study the environment industry was described as having "a dual structure with a small number of large firms accounting for about 50 per cent of output in individual market segments and a large number of smaller firms accounting for the remainder. Surveys show that about half of the environment industry is composed of small firms employing less than 50 people. However, product markets in North America, Europe and Japan are dominated by the presence of a few, large companies who compete on the basis of advanced technologies. The Japanese market is the most highly concentrated as the major suppliers are generally subsidiaries of large diversified firms who are active in several product markets.

Because of the diversity of the environment market and relatively low barriers to entry, suppliers have found numerous routes of access to the industry. It has proved fertile ground for start-ups and entrepreneurial ventures with smaller environment firms ranging from high-technology suppliers of chemicals, instruments and consultancy services to low-technology producers of recycling bins and suppliers of waste transportation services. Most small- and medium-sized environment enterprises are specialised, owner-managed and offer a limited range of equipment and services. In total, there are estimated to be some 30 000 firms in North America, 20 000 in Europe and 9 000 in Japan''. These general characteristics seem to be true for the environmental biotechnology market as well.

The size of the total environmental market was estimated at $200 billion in 1990 and growing by 5.5 per cent annually to about $300 billion in 2000 (Table 20).

However, these overall data do not show the large differences which exist between countries and between activities or trends in the four compartments (water, soil, air and solid waste). Figures 25 and 26 give some examples for actual and expected expenditures in some OECD Member countries per environmental compartment.

Equally, there can be major deviations between industrial sectors with respect to environment-related investment (Table 21), reflecting the importance of specific sectors within individual countries.

Despite these differences, it is clear that the total environment industry is substantial, both in turn-over and in employment, is trade-oriented and influences national trade balances (Table 22).

One can only estimate the role of biotechnology in the current market for environmental technology. Depending on country and effluent type, between 50 and 80 per cent of the waste water treatment facilities will be based on some form of biological treatment. For air, soil and solid waste treatment this is significantly lower and varies from 1 to 10 per cent at present. Overall, this roughly indicates that 15 to 25 per cent of the total environmental market, equalling some $40 billion annually, has a biotechnological component.

Some additional specific data are available on the total Dutch biotechnology markets, which amounted to 85 billion Gld in 1990 (Table 23). The environmental biotechnology component of 2 billion Gld equals about 20 per cent of the total Dutch environmental market of approximately 8 billion Gld. This agrees well with the indication in the previous paragraph that 15-25 per cent of the environmental market has a biotechnological component. Environmental biotechnology also accounts for 11 per cent of the biotechnological companies in the Netherlands (Figure 27). Thus, biotechnology is a serious player in the growing environment industry, and environmental technology cannot be neglected in the growing biotechnology spectrum.

Concerning the evolution of environmental markets, three distinct phases are identified. These phases are characterised by factors such as the focus of environmental policy, the capabilities of the supply side and the sophistication of the markets in terms of the environmental management strategies adopted. The progression from one stage of development to another is an evolutionary process, driven both by regulation and the technological capabilities of the countries involved.

In phase I (developing markets), there is little, if any, environmental legislation and, virtually no indigenous supply industry; the environmental management techniques used by firms, if any, will tend to be proven technology for pollution control, particularly for controlling emissions to air and water. Such techniques would primarily be acquired through licensing arrangements or direct market penetration by firms with the necessary "know how". In phase II (more developed markets), there is more wide-ranging, but non-integrated environmental legislation, with greater attention given to waste management and contaminated land issues. The indigenous supply industry is more established, with proven capabilities in certain areas, and more emphasis is placed on technological innovation as a means of addressing particular environmental problems. In phase III (mature markets), there is an integrated approach to policy developments (environmental and industrial), a very strong

Table 20. **Forecasts of market trends for the environment industry in OECD Member countries by compartment**

(US$ billion)

	1990	2000	Annual growth reate (%)
Equipment	152	220	5.0
Water Treatment	60	83	4.0
Solid Waste Management	40	63	6.4
Air pollution	30	42	4.4
Other (soil remediation and noise)	22	32	5.1
Services	48	80	7.4
Total	200	300	5.5

Source: OECD, 1992.

148

Figure 25. Expenditures on environmental technology per natural compartment in 1990-91

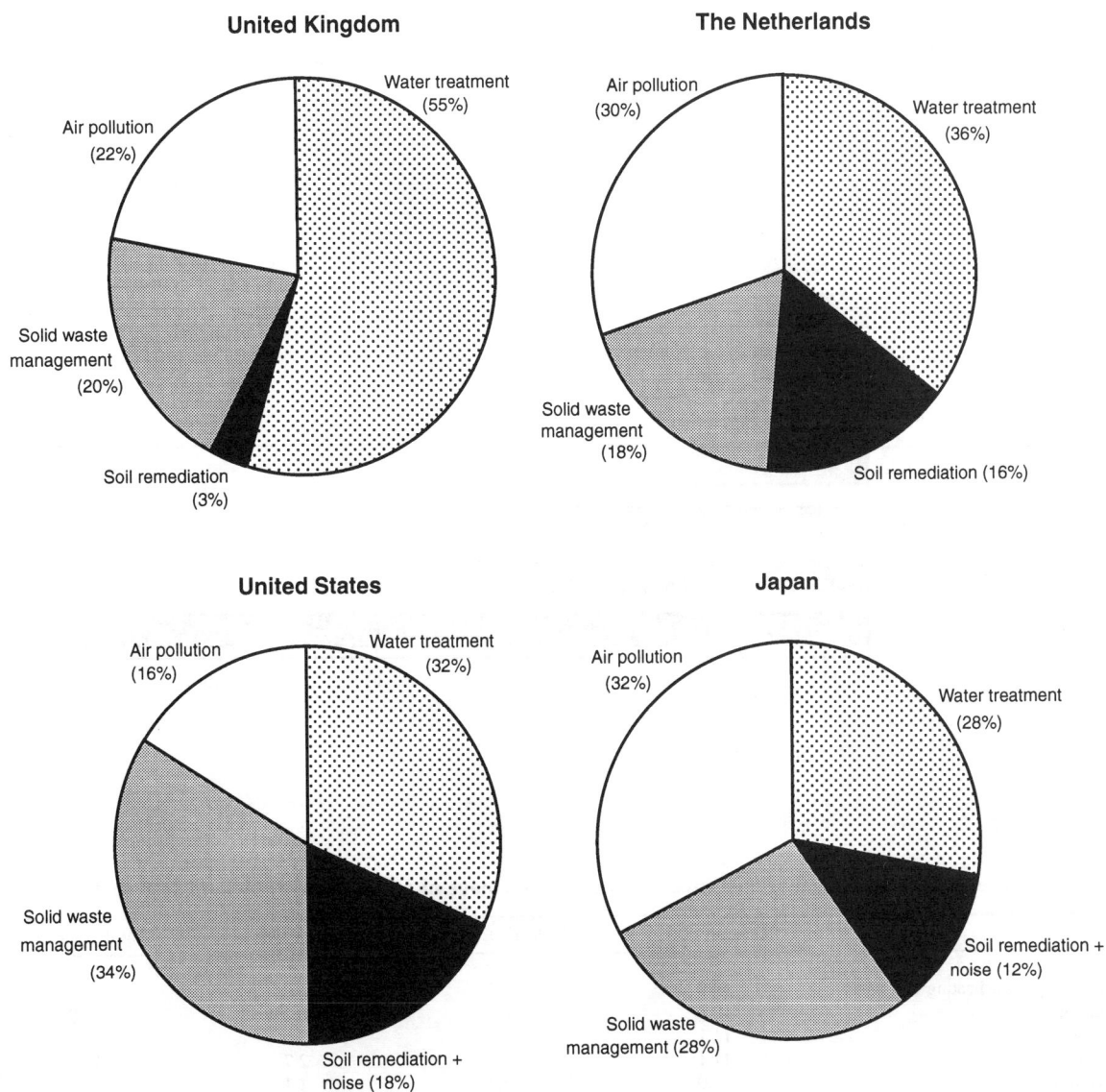

United Kingdom

- Water treatment (55%)
- Air pollution (22%)
- Solid waste management (20%)
- Soil remediation (3%)

The Netherlands

- Air pollution (30%)
- Water treatment (36%)
- Solid waste management (18%)
- Soil remediation (16%)

United States

- Air pollution (16%)
- Water treatment (32%)
- Solid waste management (34%)
- Soil remediation + noise (18%)

Japan

- Air pollution (32%)
- Water treatment (28%)
- Soil remediation + noise (12%)
- Solid waste management (28%)

Sources: VLM, 1991; CEST, 1992; OECD, 1992.

supply industry, and a focus on clean technologies and other process-related techniques as the main environmental management option.

A European map could be drawn indicating regions according to these three phases. At the present time probably only the Netherlands, Germany, Switzerland, Austria and the Scandinavian countries could claim to be mature markets. All have highly developed environmental policy frameworks, structured around some form of integrated control philosophy. All have strong supply industries which have benefited from the severe regulatory environment in their home markets to exploit opportunities, and all have an established reputation for innovation in clean technologies (including bioprocess technology) and environmental management.

Figure 26. Estimated levels of expenditure on twelve key environmental issues

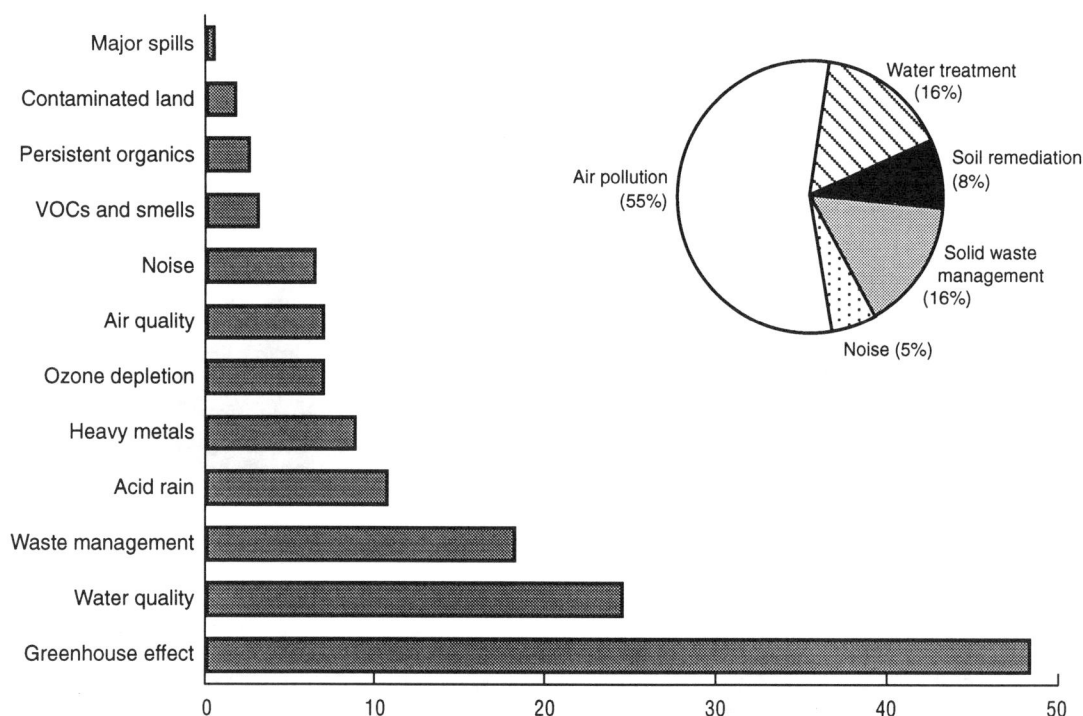

Source: CEST, *Industry and the Environment: A Strategic Overview*, 1991.

Table 21. **Environment-related investment as a percentage of all investment, by industrial sector**

Sector	Germany 1986	Netherlands 1986	Japan 1988	United States 1987
Electricity, gas, heating and water	19.8	8.8	5.8	n.a.
Mining	18.3	2.1	6.2	2.0
Paper	11.1	2.8	3.0	5.0
Steel	11.1	2.7	2.7	11.3
Refining	10.0	22.2	1.1	7.0
Chemical industry	9.4	3.5	1.3	5.2
Metals, first processing	8.3	2.0	1.1	10.6
Foundries	7.5	n.a.	n.a.	n.a.
Mineral processing	5.9	2.7	n.a.	n.a.
Leather	4.3	4.9	n.a.	n.a.
Motor vehicles	3.3	0.2	1.0	3.0
Food	2.8	1.8	n.a.	2.5
Textile	2.6	0.7	2.2	1.8
Printing	1.7	0.1	n.a.	n.a.
Mechanical engineering	1.3	5.0	0.7	n.a.
Electrical/electronic engineering	1.2	0.3	1.3	1.8
Construction	1.1	1.8	1.6	n.a.
Machine tools	0.9	0.4	7.0	0.7

n.a.: Not available.
Sources: Sema Group 89; Statistisches Bundesamt, Germany; Centraal Bureau voor de Statistiek, the Netherlands; Department of Commerce, United States; Environment Agency, Japan.

Table 22. **Production, employment and trade estimates for the enviroment industry in OECD Member countries**

	Production (US$ bn)	Employment (× 1 000)	Production exported (%)	Trade (US$ bn)
United States	80.0	800	10	4 000
Canada	6.0	50	n.a	Negative
Europe	(68.0)	(600)	(20)	(8 000)
Germany	27.0	250	40	10 000
France	12.0	90	14	500
United Kingdom	9.0	75	17	500
Italy	5.0	40	n.a	Negative
Other	15.0	145	n.a	n.a
Japan	30.0	200	6	3 000
Other	1.0	50	–	–
Total	185.0	1 700	–	–

n.a: Not available.
Source: P. Hesselink, TNO, Delft, Netherlands, based on OECD data, 1990.

Less favourable prospects in the domestic market in countries with a mature market may stimulate companies to lay greater emphasis on exports. Companies selling bio-environmental products in countries like Germany and Switzerland already export a substantial amount of their products to companies in Europe as well as other parts of the world such as Asia and the Middle East. In general, it is expected that countries in the south of Europe will gradually take a greater interest in environmental problems, as a result of which exports of (bio)-environmental products to these countries may be expected to increase. A number of companies are also endeavouring to build up commercial contacts in east European countries. This is far from easy because of the transitional problems of these states.

The majority of markets (including the bio-environmental market) in Western Europe (United Kingdom, France, Italy, etc.) are classified as more developed markets (phase II), although some would obviously be more mature than others. All countries have fairly comprehensive, but non-integrated environmental policies, indigenous supply industries with established reputations in certain areas, but with the primary focus still being on curative (add-on) control techniques as the main environmental management option. The same seems to be true for Japan and the United States, which both have a mature environment industry in some sectors. In the general water treatment area, the market is already mature for products which current Japanese environmental protection laws and regulations mandate. In other areas, however, environmental biotechnologies are only in the embryonic stage and provide considerable scope for future growth.

The situation in some southern and east European countries is embryonic. In most cases there is virtually no environmental legislation and/or no effective enforcement. This in turn has given rise to severe environmental problems, but without legislation there has been no rationale for an environmental supply industry. The countries in east Europe in particular have to make a drastic jump in environmental standards, and are looking to the West both for the expertise and the resources to assist them in making this transition.

Table 23. **Estimated Dutch production with a biotechnological component
in industrial sectors**

Industrial sector	Estimated production (in billion Dutch guilders)
Food and feeds	28.6
Chemicals	28.9
Health care	1.5
Agriculture	4.9
Environment	2.0
Hardware & engineering	20.0

Source: Netherlands Ministry of Economic Affairs, 1990.

Figure 27. **Distribution of Dutch companies in biotechnological application areas**

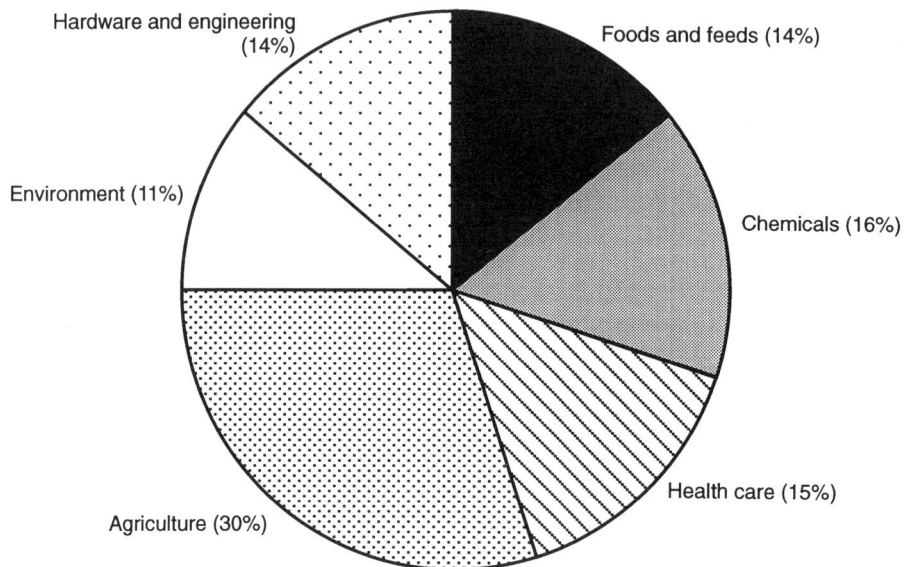

Hardware and engineering (14%)

Environment (11%)

Agriculture (30%)

Foods and feeds (14%)

Chemicals (16%)

Health care (15%)

Source: Ministry of Economic Affairs,The Netherlands, 1990.

3. Turnover and trends in environmental biotechnological market development

Air/off-gas

Demand side: market size and dynamics

In general the activities in the air/off-gas treatment segment are growing. Up to the year 2000 the growth rate in Europe is expected to be about 15 per cent per annum. A major driving force behind this dynamic market is the EC-wide abatement of SO_2 emissions, resulting from the directive on large combustion units. Coming EC directives are likely to be a strong incentive for growth of the off-gas treatment sector, including biotechniques. In the Netherlands, companies in this field suffered losses in 1991 but a sharp rise in sales is expected in the coming years.

The current impact of biotechnology in this segment is rather limited, but undoubtedly increasing, particularly if fuel desulphurisation techniques are developed and biotechniques gain more market acceptance. In this segment, about 20 per cent of the products are biotechnological, being mainly biofilters and bioscrubbers. However, as the costs per product are relatively low, the economic impact of biotechnology in this sector is lower as well. As an example, the Dutch turn-over of both products was $17 million in 1991, of which 90 to 95 per cent was for conventional biofilters.

All over Europe, the investments in air pollution control are expected to grow from $4.0 billion to $5.4 billion through 1995. With respect to size, Germany and the Netherlands are leading, owing to their stringent legislation in this area. In 1987, about 70 per cent of the investment for air/off-gas treatment in Europe was made in Germany. In the Netherlands, the two largest companies in this area are also important players on the world market. In addition, there are five or six smaller companies.

152

Supply side: companies

In Europe there are more than 500 competing companies, including approximately 50 with activities in biological air/off-gas treatment. These are mainly engineering companies, offering equipment to reduce gaseous pollutants. Other companies are trying to develop systems to remove the polluting elements from fuels before combustion. A major area of interest is removal of sulphur from coal and, to a lesser extent, from oil. However, this field is still in an embryonic stage with respect to biotechnological solutions. More progress has been made in the replacement of environmentally harmful organic solvents by water-based solvents. This offers excellent opportunities for biotechniques as these emissions are usually biodegradable.

The main basis for competition is compliance with governmental standards, followed by price of the equipment, maintenance/operating costs and reliability. Key factors for success are technological and engineering know-how, particularly the ability to deliver turnkey projects. However, one of the most important factors for future success in this subsector is the integration of skills in biotechnology with engineering and hardware design capabilities.

Competition

With the exception of biological gas treatment, competition is relatively intense, as several heavy engineering companies have chosen this market as an area of diversification. The character of the competition in this segment of the environmental biotechnology market in the Netherlands is oligopolistic with two companies that are also active on the world market (especially in the United States) in biofilters.

It seems rather easy to enter the market because of the seemingly simple techniques used in biofiltration: a closed box with organic material. Achieving a market share is nevertheless a problem because of the poor quality and reliability of these over-simple biofilters.

There are a number of small start-up companies, but most are unsuccessful because, with their lack of know-how, they are not capable of developing new techniques or do not have close working relations with research institutes. Another selection mechanism is the "no cure no pay" principle, which means that these small companies have to take enormous financial risks. Learning processes and (venture) capital needed for R&D can be barriers to achieving a reasonable share in the market for air/off-gas treatment. Subcontractors play no significant role in this market.

Water treatment

Demand side: market size and dynamics

The total annual investment in water purification and waste water treatment plants in Europe was $11 billion (1987) with total operation expenditures being close to $21 billion in 1990. In Europe, about 60 per cent of the water used in households (100 litres per day per capita on average) is treated in waste water treatment units, while in the United States and Japan about 70 per cent and 30-40 per cent respectively of water used in households is treated. This total waste water treatment market was estimated to be $19 billion for the United States and $5 billion for Japan in 1990. The large majority of installations are aerobic. Anaerobic waste water treatment is a very promising technology in this segment, because it can satisfy the more severe legislation and is an appropriate technology for both the "replacement" as well as the "innovation" market. Its impact is growing although still limited.

In general, the waste water treatment industry is growing but maturing. Its degree of maturity varies widely depending on the individual country considered. In countries like the Netherlands and Germany, the industry is approaching full maturity. The Dutch market for domestic and industrial waste water treatment is reaching saturation for aerobic as well as anaerobic installations as a result of strict regulation introduced in 1970 (Wet Vervuiling Oppervlaktewater). In 1991, the total sales of aerobic and anaerobic waste water treatment systems in the Netherlands was approximately $100 million out of approximately $180 million total sales for waste water treatment.

In countries such as Italy, France, Spain and Ireland, where a substantial proportion of waste water is still discharged without any treatment into the sea and rivers, the greater part of the growth still has to take place. Despite this, France especially has some highly developed suppliers of equipment and services.

Municipal waste water treatment units are mainly mechanical-biological units operated by governmental agencies. Industrial waste water (which constitutes about 60 per cent of the total waste water in the Netherlands, for example) is treated in privately owned waste water treatment units which are mainly mechanical (50 per cent), physico-chemical (25 per cent) or biological (25 per cent). However, proportions vary widely depending on the particular country considered and government policies (which largely determine whether operating one's own treatment plant or relying on a municipal or central sewage system is more appropriate).

Supply side: companies

In this huge and state-guaranteed market a large number of companies are involved. While there are limited changes in the scope of applications, the number of competitors is relatively stable. As an example, western Europe has about 350 major suppliers of waste water treatment units.

As a consequence of relatively early legislation and innovation, the Netherlands overall market growth has a strong position in the world market for these systems, with two companies as important players in anaerobic waste water treatment systems. These Dutch companies will achieve by exporting their installations world-wide. They compete on the world market by building up a licensee network; they have a number of patents and a practical know-how base that have provided them the opportunity to build up this world-wide network. This was stimulated by the fact that most countries, including developing countries are nowadays heavily investing in waste water treatment for environmental or hygienic purposes.

In Germany, about 50 companies are active in the market for waste water treatment. The increasing environmental legislation in Germany guarantees a growing market for these companies.

A distinction should be made between construction companies and the specialised environmental biotechnology companies. For the former group there is merely a "replacement market", leaving only a small number of buyers who will eventually be obliged to treat their waste water. For the specialised environmental biotechnology companies a growing market of dedicated systems has appeared, as a consequence of stricter legislation.

Several companies are active in the field of innovative biotechnological R&D for water treatment, both for in-house purposes and for commercial activities. However, the innovative biotechnology content appears to be relatively limited as engineering skills still play a key role. Thus, biotechnology R&D is limited largely to sludge properties and quantities, the selection of the most appropriate strains and performance and stability of the installation from a biotechnological point of view.

The market for projects over $1 million has an oligopolistic character. In this segment there are companies that supply "turn-key" systems to end users: municipal authorities for domestic waste water systems and industrial companies for industrial waste water treatment. The companies in question seek contractors/builders to build the installations but remain responsible for the whole project.

Competition

The most important competitive factor is the price of the equipment. The performance of the waste water treatment units is usually sufficient for most standard purposes and applications. Total service (including complete financial package and operating assistance) is also increasingly important.

The market for waste water treatment in countries with an established legislative system is relatively saturated. Competition between companies supplying equipment, chemicals and service to water treatment units is consequently severe. By contrast, in those countries currently implementing legislation, many opportunities exist but there will be strong competition from internationally operating companies for start-up companies within the country.

Competition is not between environmental biotechnology and other environmental technologies but between the suppliers of biotreatment techniques. A number of companies escaped from this heavy competition, and found a specific group of clients and market niches such as slaughter houses and breweries or by offering waste water treatment units as part of a total package. In most cases, the supplier already had good contacts with the potential buyer as a consequence of activities in other areas. Because of existing relations new deals are made very easily. An important aspect is the small number of contractors the buyer has to deal with: the supplier builds the installation and also delivers the additional environmental equipment.

The factors for success are:

– cost;
– technological and engineering know-how (ability to develop and offer tailor-made turn-key systems);
– company image (through references, advertising and promotion);
– technical services and applications;
– development capabilities;
– links with end users and authorities and product/service range.

Entering the market for anaerobic waste water treatment units is very difficult; experience, the availability of capital and references are as important as having highly qualified personnel.

Soil treatment

Demand side: market size and dynamics

In western Europe the process of inventorying old sites has been initiated. Some countries such as Germany and the Netherlands have already estimated how many sites need to be restored. In Germany, there are 120 000 sites of which 10 per cent need to be treated in the short term at a total cost of $15-20 billion. In the Netherlands, there are about 6 000 registered severely polluted sites of which about 900 need to be cleaned up urgently. The total costs of cleaning the severely polluted sites is estimated to be around $30-60 billion in the Netherlands alone, based on conventional techniques and existing legislative requirements. Decades will be required for clean-up.

According to recent estimates, up to 60 000 sites in Europe will have to be cleaned urgently, representing a yearly market of many billions of dollars up to the year 2000. The ultimate numbers are likely to be considerably higher, as more countries embark on making an inventory of contaminated sites, particularly if common legislation is adopted and enforced at the EC level.

In contrast to the above figures, the actual market for soil treatment and land reclamation is much smaller. Table 24 estimates the annual turnover in this sector for 1990 and 2000. It is clear that this market has only a fraction of the size it could have. Therefore substantial growth can be expected in the coming years (see below). It should be remembered that great fluctuations between individual countries will occur and that trends can be substantially influenced by national legislation, priority-setting, and funding.

The soil remediation market can be divided in two segments: contaminated soil and groundwater. In the first segment no more then 10 per cent of the total amount of treated soil (in the Netherlands: about 100 000 tonnes) is treated biologically at present (landfarming and bioreactors). In the second segment, the contribution of biological techniques is larger, thanks to the use of aerobic and anaerobic water treatment techniques. In this case the subsoil water is extracted, treated and reinjected or discharged.

The activities in the site clean-up segment are growing; annual market growth rates for the period 1990-95 are expected to exceed 10 per cent. Key driving forces behind this substantial growth projection are the identification of toxic waste disposal sites as a priority area by public opinion, the availability of increasing funding, and the reliance on more sophisticated technological solutions. Quick and simple techniques such as landfilling, export to less developed countries or sea-dumping become more and more restricted. Eastern Europe is one of the most important potential growing markets for soil remediation companies, as long as it stays a priority for funding from organisations such as the EC or the World Bank.

After a period of enormous growth in market turnover in the Netherlands, the soil remediation market has stagnated, mainly as a result of a collapse in demand and a lack of clarity and certainty in the governmental criteria for clean soil. In Germany, demand has stabilised while the number of competing firms has grown. The unification of Germany has led to a paradoxical situation. On the one hand, there are many contaminated sites in the eastern part, on the other hand the reconstruction of the economy requires huge governmental expenditures, as a result of which soil treatment has a lower priority.

For site clean-up and reclamation, the growth rate for construction engineering will not exceed that of overall GNP. However, this sector will continue to amount to approximately one-third of the market volume. The number of participants supplying technology is generally stable (large companies). The growth rate for process engineering, on the other hand, is expected to be high, especially for treatment of industrial sites and landfills. Biological treatment is set to capture a significant share of total market volume by the year 2000. The product line of the companies active in this field is typically narrow and often tailored to satisfy specific needs.

Table 24. **Annual market size for soil remediation in the years 1990 and 2000**
(billion dolllars)

	1990	2000*
EC	3.5	6.0-9.0
Japan	1.5	3.0-4.0
United States	6	10.5-15.5
OECD total	10	19.5-28.5

* Estimates based on OECD and EC growth figures for specific sectors.
Source: OECD, 1992; EC, 1992.

The current impact of biotechnology in this sector is marginal. Overall, only a minute fraction (about $50 million) of the total existing market for site reclamation, estimated at about $5-10 billion world-wide, derives from biological treatment. The projected growth rate for biological treatment for site reclamation is expected to be 5-10 per cent annually through 1995, starting from this very low base.

Supply side: companies

Most of the companies in this market segment originally started their activities in the field of municipal solid waste collection or road and infrastructure construction. They subsequently identified contaminated soil as a large and lucrative market requiring the type of equipment and scale of thinking they were used to. In Europe, industry participants are mainly small companies, but large chemical and energy companies are now entering this business as well. Their number has increased rapidly in the last two years.

In the Netherlands, four large companies and 20 smaller companies are active in this field. The larger companies are active in both sectors and soil treatment is only part of their total activities. Although the Dutch companies are pioneers in microbial soil treatment on the world market, export has started only recently. In the Netherlands there is a potential for considerable demand for soil treatment services, especially biological options. These latter techniques have a high environmental efficiency and are relatively cheap (see Chapter 7). However the effective demand is rather low, due to the policy of the Dutch Ministry for the Environment (VROM), which holds to the so-called A-norm for the end product: close to a 100 per cent clean and multifunctional soil.

Most companies in this field are private. Major exceptions are in Denmark and Sweden, where toxic waste management including contaminated soil is centralised and operated by a state owned agency. In Germany, the Bundesländer (federal states) have a share in some companies. In Germany, about 100 companies are offering equipment or services but only approximately 30 companies offer the whole service range including biotechniques.

Over the past few years, several soil remediation companies have been established in this field, attracted by the growing market size and the relatively modest entry barriers. However, as a natural process but also as a consequence of economic recession in the early 1990s, a trend towards consolidation is visible, with large companies swallowing smaller industry participants. Some companies are even leaving this market segment as it becomes tightly regulated and entry barriers increase. As a result, only a few large groups in a position to offer a wide and complete range of services to industry are likely to emerge. On a long-term basis they are able to guarantee a strict compliance to environmental standards and have the resources to built increasingly expensive units (mainly for incineration at present).

Competition

Competition in the soil reclamation market in countries with strict legislation is rather severe and will increase in the future. In the Netherlands, due to uncertainty in legislation and enforcement, over-capacity is a severe problem: companies in this segment, which specialise in all kinds of *ex situ* treatment, digest approximately 100 000 tonnes a year, while their capacity is approximately 1 million tonnes/year. This has resulted in a situation of sharp price-competition, big differences in the quality of the offered product, and even the importation of polluted soil from other countries to keep the installations going and to create a small return on investment.

Entering the market is not very difficult. However, it is more difficult to remain. For example, in Germany, every time there is new order, at least five companies of the top ten will be at the negotiating table.

A problem for bioremediation is that several projects have been carried out throughout Europe by inexperienced firms often selling a "biological miracle": the low quality and results of such projects are a major threat to further implementation of serious environmental biotechnology. Newcomers are often big companies which repeatedly attempt to enter the market by underbidding, with the purpose of having a project that may serve as a reference in their further efforts to solicit new clients with more profitable projects.

One of the results of a survey carried out by three big Dutch companies active in this market is that a newcomer company wanting to deliver high quality services and not just using classical open landfarming techniques, will have a hard time surviving because of its need for a more expensive high- quality approach and services. Other barriers for newcomers are the lack of references and the "vergunningenbeleid" (permit policy). The established companies hold to the opinion that the highest quality of soil reclamation can only be produced by the bigger companies and their subsidiaries.

Especially in the present financial climate, when goverments and industries have limited money, the capacity of the soil remediation companies is much higher than the actual demand for clean-up. This demand is not always recognised, as some oil companies, for example, have surrounded their contaminated soil problems with secrecy, and the need for clean-up appears smaller than it is in reality.

Although in this market segment price competition is the most important competition factor, the sale of cleaned or "clean" soil is one of the biggest problems for the companies and is now almost as important a competition factor. The bigger companies in this sector try to secure their market for clean soil, through contracts with organisations that use a lot of soil (harbour companies, the State Water Authorities or the local public service for public gardens) or by acquisition of "landfill" locations. Where there is overcapacity, there is severe price competition which may appear as competition between technologies. Where landfilling is still permitted this is inevitably the cheapest route of disposal and thus aggravates the problem of too much treatment capacity.

Overall intensity is moderate but could increase as price becomes more important as a basis of competition. Typically, competitive intensity is higher in the civil engineering segment compared to the process treatment segments such as incineration, whose entry barriers are higher because of the need to construct expensive equipment. Performance and compliance with the governmental standards represent another important element in the basis of competition.

Key success factors for companies in the soil treatment subsector are:

– technical know-how (including ability to deliver turnkey systems);
– close link with authorities;
– cost structure;
– tight control over the internal organisation;
– image (references, reputation);
– service (to relieve the consumer completely from any trouble deriving from waste disposal operations);
– range of services offered;
– applications development/links with end users.

Solid waste

Demand side: market size and dynamics

This industry includes all services and equipment used to treat and dispose of solid waste, irrespective of its origins. Table 25 indicates market size and growth for some OECD Member countries.

The market for solid waste treatment in Europe can be estimated to be about $8 billion (in 1989) growing to $17 billion by 1995 (including $2 and 3 billion respectively for investments in fixed assets). Approximately 2 billion tonnes of solid waste are produced in the EC every year. Of this, about 160 million tonnes consist of industrial waste and 20-30 million tonnes are toxic waste.

The average waste composition in Europe can be roughly estimated (Table 26).

The type of solid waste treatment varies widely depending on the various geographical regions and countries (Table 27).

About 660 incineration units exist in Western Europe. This number is expected to grow to 850 over the next ten years. The number of composting and sorting units in Western Europe is expected to grow from 160 in 1986 to about 550 in the year 2000.

Table 25. **Annual market size for solid waste treatments in the years 1990 and 2000**

($ billion)

	1990	2000*
EC	8.0	12.0-25.0
Japan	5.5	8.5-17.0
United States	19.5	28.5-60.5
OECD total	34.5	51.5-160.0

* Estimates based on OECD and EC growth figures for specific sectors.
Sources: OECD, 1992; EC, 1992.

Table 26. **The solid waste compositions in Europe**

Component	Share (%)
Paper	30
Glass	10
Metal	2-10
Plastic	10-15
Other	35-58
(Organic)	(60-80)

Source: EC, 1992.

The Dutch government has decided that all vegetable, fruit and garden waste (GFT-waste) must be sorted and separately treated from 1994 onwards. This has created large perspectives for new composting companies. It is expected that in 1994, one million tonne of GFT-waste will be collected in the Netherlands, a tripling of the amount in 1992. Consequently, at present more than seven companies are developing composting installations.

Companies supplying aerobic and anaerobic composting units in countries with a tight environmental legislation system are in severe competition. The question is who will be the first to develop a composting unit which can digest these vast amounts of GFT-waste in an effective way at low costs per tonne. Competition from other techniques is not apparent. However, to what extent and how this potential market will become reality is heavily dependent on the costs of treatment and of the size of the market for the compost endproduct.

Another environmental problem is the enormous surplus of animal manure. This has generated a large demand for manure treatment installations. However, because of the long-term technological and economic problems connected to the development of these installations, the effective market is still small.

Concerning the overall dynamics, the solid waste treatment industry is in a growth phase. Biological treatment of sewage sludge is increasing and a market for biological installations is already substantial and will most probably increase in the future. At the moment biological treatment processes are developed for the treatment of harbour and dredging sludge. The market for these kinds of processes and installations is, given reasonable cost efficiency, guaranteed, due to the enormous amounts of sludge produced.

Supply side: companies

Solid waste treatment is performed by both private and state-owned companies. Often the two groups of companies co-operate. There are many small regional companies in Europe dealing with transport and collection. Most do not have their own treatment units and rely mainly on landfill.

The trend in this sector is towards company concentration as governmental regulations become tighter public scrutiny puts pressure on the industry and low capital-intensity solutions, such as landfill disposal and "pirate" discharges, are no longer feasible. This trend, already noted in the United States, is likely to accelerate the disappearance of the smaller industry participants and the emergence of "industrial majors" in Europe (such as: Laidlaw-Waste Management and Browning-Ferries Industries in the United States).

Table 27. **Share of three types of solid waste treatment for five European countries and western Europe as a whole**

(in %)

	Incineration	Composting	Landfill and other
Western Europe	40	10	50
Austria	18	22	60
Netherlands	3	17	80
Sweden	50	10	40
Switzerland	80	2	18
Italy	6	4	90

Source: EC, 1992.

Competition

Price, service and compliance with existing regulations represent the basis of competition. Key success factors in the solid waste treatment sector are:

– technological know-how (in order to comply with the increasingly tight regulations);
– range of products/services offered;
– applications developed/technical services;
– financial power (to finance the new hardware requirements, and possibly face legal claims);
– cost structure;
– image and relations with authorities.

Overall the competitive intensity is moderate. For landfill disposal and mechanical treatment the competitive intensity is high, while it is lower for thermal or biological treatment.

Chapter 9

Supplier Company Policies

1. Five model companies in environmental biotechnology

Studies of the environmental biotechnology sector show that the structure of the environmental market in Europe, the United States and to a lesser extent Japan is complex and highly fragmented. The supply side has a great diversity of companies active in the field of environmental biotechnology. This diversity results from differences in company size, structure and products. There are a large number of mainly small and specialised firms, most of which belong to a large international firm or co-operate with other companies, universities or research institutes in order to survive in an increasingly competitive arena.

Depending on the definition of environmental activities, between 6 500 and 20 000 environmental protection and waste management firms are active in Europe. Of these, two-thirds have annual turnovers of less than ECU 5 million (approximately $6 million) and almost three-quarters are owner-managed. The same is valid for the estimated 38 000 firms in North America and to a lesser extent the 9 000 firms in Japan. This small average size implies that most firms are unlikely to benefit directly from large-scale operation. They will mainly play the role of specialised subcontractor or execute small and medium-sized contracts in their home region. They are likely to lack the resources and reputation needed to compete in increasingly competitive international markets. This holds especially for the deregulated European market unless the quality and marketing of these relatively small companies is outstanding. Otherwise, their activities will be limited with respect to geography and activities. A rationalisation process of acquisition and mergers is taking place at present, and environmental biotechnology firms are part of this process.

In Chapter 8, four separate compartments of the market for environmental biotechnology were discussed. In each compartment a heterogeneous group of companies is active. Five basic types of companies have been identified based on the results of the interviews with representatives of individual companies. These types represent most of the companies in this sector. Generally, the individual companies will have most of the characteristics of one of these models. As reality is more complex, however, it is not uncommon that companies in this sector have characteristics of more than one model.

Model I companies are big multinational companies with main activities in the chemical, pharmaceutical, agricultural and/or food sector. These large companies have subsidiaries or departments with biotechnology interests. This group includes companies whose activities are based on traditional fermentation processes. Most of these companies have environmental policies and statements. Sustainability in production and development is more and more an integral part of company policy.

This type of company buys environmental products and services, but also develops environmental biotechnology products and services for its internal market. In some cases these activities have become so successful that they have become a profit centre of their own. Depending on a number of considerations (profitability, public image, core business, etc.) these new business units will be kept in house or sold.

Model II companies are small, innovative and dedicated environmental biotechnology companies. New companies of this type are created specifically for business activities using environmental biotechnologies. In some cases, environmental biotechnology R&D is their only business. They have strong relations with public R&D institutes such as universities and R&D organisations. Often they originate from R&D institutes or from model I companies in order to commercialise R&D results and patents (Box 1).

Model II companies are relatively high-technology companies and concentrate on niche markets such as development of specialised treatment methods, specialised microbial strains or analyses of specific toxic contaminants.

Model III companies are equipment and installation, construction, and manufacturing companies. Their core business is development, manufacturing and installation of process equipment or a combination of these three activities. They have also developed or adapted specific equipment for waste water treatment and air/off-gas treatment. Their environmental biotechnology activities are located in a department or even in a subsidiary of a large engineering company, most likely specialised in the field of environmental systems and equipment. These companies are now active in the air, water and solid waste segment and often add development and consultancy to their package. Biotechnology has been adopted with a view to enhancing the selection of environmental products (Box 2).

Model IV companies are subsidiaries of large building and infrastructure companies which embarked on the remediation of contaminated soil. These large building companies offer construction engineering technologies like barrier technology, soil stabilisation and earth moving. Because of the type of equipment they traditionally use, they fit very well in the new environmental market, mainly in the soil remediation area. These Model IV companies develop their own equipment for providing soil treatment services to third parties. They might also become active in solid waste treatment.

Model V companies are consultancy bureaus/companies, the environment being one of their fields of specialisation. The company is able to advise industrial companies or governments on how to solve their waste and effluent problems, environmental biotechnology being one of the potential solutions.

Apart from these five, more or less prototype companies, a broad range of other companies carry out various support functions, ranging from raising venture capital, through provision of services (information, consultancy, construction planning) to provision of equipment, active substances (microbes, chemicals, enzymes) or a combination of these.

For each of the five model companies, characteristics are presented below in relation to R&D expenditures, R&D co-operation and networks, patent position and strategy, company strategy and attitude towards biotechnology.

2.   R&D expenditures

Many companies in this field are committed to R&D but most of them only use R&D to support the products which they already sell. Far fewer companies have R&D as a product on its own. New demands for sustainable and process-integrated environmental solutions require new biotechnology activities in basic research and industry, such as the selection and use of new microbial strains, improved methods of analysis and novel biotreatment methods. However, the environmental biotechnological industry is characterised by medium value-added products and low to high value-added consultancy services. This implies that companies delivering products in environmental biotechnology hesitate to allocate substantial budgets to R&D because of their relatively low margins and extended periods for return on investment. This could become a serious barrier to implementation and innovation as most of the companies in environmental biotechnology will depend on products. Therefore, incentives for R&D, for example subsidies and sponsoring of demonstration or accelerated innovation, will be important in environmental biotechnology.

For most of the Model I companies, environmental biotechnology is not part of their core business. So although R&D expenditures on biotechnology might be high, environmental biotechnology is not or rarely included. However, there are exceptions. Some of the multinationals have activities in environmental biotechnology as a spin-off of their core-business, by inventions or serendipity. In this case the add-on technologies are mainly in waste water and off-gas treatment.

This might change substantially in the future, however, as the environmental response is becoming an integrated part of company business. Model I companies especially are expending more and more effort on developing processes which are less polluting or inherently clean. One of the most important means to do so is by using biotechnological techniques such as biocatalysis in their production processes. For this reason, these companies are becoming increasingly convinced of the advantages of bioprocessing: less water and energy consumption and less waste. The costs for developing new biotechnological production methods are then part of the core-business and also regarded as an integral part of the overall company R&D expenditure. Although most companies could give an indication of this expenditure, they did not for confidential reasons. Depending on the type of company and market, an estimate of one to ten percent of the total R&D budget allocated to biotechnology seems reasonable.

Model II companies, the dedicated environmental companies, spend a large percentage of their annual turnover on R&D. For companies whose core business is environmental biotechnology this percentage can rise to 75 per cent of the company's turnover, mainly as contract R&D on behalf of third parties. In the innovative companies that offer products and services to the market, the R&D expenditure may be as much as 10 to 15 per cent of the total annual turnover of the company.

Model III and Model IV companies are in most cases large companies. In general their environmental biotechnology activities are very small. However, there are exceptions, and some companies have developed specialised treatment equipment, mostly in co-operation with universities or public research institutes. This gives a local focus of environmental biotechnology R&D activities in these companies. In some cases Model III and Model IV companies acquire biotechnological expertise by hiring a (very small) biotechnological staff and/or by co-operation agreements with universities or R&D institutes. On the whole, companies of both Models III and IV spend a very low percentage of their annual turnover on environmental biotechnology R&D, generally less than 1 per cent. Mergers with and acquisition of specialised environmental biotechnology firms (*e.g.* Model II companies) are aspects of their strategic options to obtain innovative knowledge and expertise.

The same low percentage of R&D expenditure applies to the Model V companies, although some of them might also have developed specialised treatment processes or equipment. In a number of cases there are no environmental R&D activities at all. This applies especially in the consultancy firms whose only investments are usually made in buying literature and software.

3. Co-operation in business

Companies operating in the market for environmental biotechnology often try to strengthen their position by establishing relationships with other companies. These companies create special forms of co-operation with their suppliers and customers in a number of ways and also have co-production and licence contracts which may be important. Joint ventures may be embarked on, most frequently abroad. In all four natural compartments of the environment industry market it is an old tradition to work closely together with suppliers, customers, engineering and consultancy firms in so-called vertical co-operation and integration. Companies work closely together in all phases of the process from the outset to the moment of completing a project. In the soil remediation market, companies will even work together with their direct competitors for reasons of minimising risks or costs. Smaller companies which are geographically closer to a polluted site are hired by larger companies. The small company cleans the soil using the biotechnological expertise of the large company. Here, subcontracting is very important.

Model II biotechnology companies are also often hired as subcontractor to provide specialist advice and expertise on parts of a project, develop a treatment strategy or deliver biotechnological products like inocula, equipment, or analytical facilities. Through such activities, Model II companies improve their flexibility, image, and network; co-operation is extremely important for these types of environmental biotechnology firms.

In general, companies producing treatment equipment (including Model III and Model IV companies) do not work together with each other on a national level; there is no horizontal co-operation. This is so because they do not want to share the technological knowledge, for which they have paid high R&D costs, with their direct competitors. Nevertheless, the situation is quite different on an international level. By establishing joint ventures and by licensing, these companies try to get a part of the market in other countries and make use of the network and the image of a local company. In exchange the "national" company offers technical expertise and familiarity with the local way of doing business.

4. R&D co-operation and the role of the research institution/industry network

Many companies carefully maintain linkages with universities and research organisations. These linkages support a network of informal contacts among entrepreneurs and academic or non-academic researchers. Through these contacts the companies gain access to recent scientific knowledge. At the same time, co-operating with universities and research organisations commonly enhances a company's reputation.

All of the Model I and Model II companies which were interviewed have R&D co-operation agreements with other companies, universities, or research institutes both on a national and international level. There are two main reasons mentioned for co-operation with other partners in the field of R&D. The first is access to the expertise of the partners and the knowledge transfer which can take place in the process of co-operation and the second is the (informal) network companies are entering. The network was especially stressed as an important way to keep informed on new technologies. An important and positive role in international co-operation and networking in Europe is played by the various EC R&D or demonstration type of programmes. Other possible reasons for co-operation mentioned in the interviews (exchange of personnel, cost savings) were rarely indicated by these companies as being important.

Most Model III, Model IV, and Model V companies spend only a small percentage of their annual turnover on environmental R&D. A number of different strategies for R&D co-operation can be identified. In some cases there are close contacts with universities or research institutes for the development of new processes or equipment. However, in most cases these companies perform no R&D and do not co-operate.

One Model III company mentioned that the research network is important for the firm, without being crucial to its performance. Their research networks makes a moderate contribution to the transfer of knowledge. The research network is very relevant with regard to informal contacts.

The network of contacts with universities and research institutes was perceived by another Model III company as important for a number of reasons. It has an important public relations effect: co-operation with a university professor or well-known expert from an institute increases the firm's credibility. The ties with universities and R&D organisations also imply that there is no need for the firm to do fundamental research themselves. This fits well in the general strategy of these Model III companies which rather prefer to hire R&D capacity for non-core business activities. In this way the firm gets relatively easy access to new developments.

The universities and research institutes also offer an opportunity to have new findings tested by a neutral third party without incurring excessive costs. The network therefore makes a substantial contribution to the saving of capital costs of transfer of knowledge.

5. Patent position and strategy

In general, there are only a small number of patents in the field of environmental biotechnology. Table 28 gives an overview of the patents of OECD Member countries in the environmental area. It shows that great differences exist among the countries mentioned with respect to patent strategy. The 1991-92 scan indicated that patenting is especially favoured in Germany and Japan, whereas it is less pronounced in the United, States, France, and the United Kingdom. In nearly all countries, most patents refer to the treatment of so-called liquids. This seems to reflect the international maturity of the water treatment market.

Table 28. **OECD Member countries environment patents (1991-92).**
Breakdown by type of pollution control

	Total	Air	Liquids	Solids
EC	491	177	215	99
Belgium	3	0	1	2
Denmark	3	1	2	0
France	20	4	10	6
Germany	212	74	98	40
Italy	6	2	2	2
Netherlands	5	3	2	0
Portugal	8	4	2	2
Spain	22	11	7	4
United Kingdom	38	10	18	10
European patents	174	68	73	33
Austria	29	6	14	9
Switzerland	5	2	2	1
Japan	306	67	122	117
USA	131	37	53	41
Total	962	289	406	267

Source: Derwent Publications, Technical and Online Services, London, 1993.

The strategy of most Model I and Model II companies with respect to these patents does not differ from their general strategy: if it is their core business they only license patents to companies that are not their direct competitors and/or that offer them an entry to foreign markets. Alternatively, patents are considered as a way of promoting themselves as experts, as having a patent implies that one really knows what one is talking about. Thus, it is a means of publicity with respect to innovative scientific potential.

On the basis of the interviews it was rarely possible to identify a specific or unique strategy for the environmental biotechnology companies with respect to patents in this field differing from their general patent strategies in other fields. The only exception is that the reluctance to patent is somewhat higher (in some countries) as it is believed that it is relatively easy to circumvent environmental biotechnology patents which deal with open microbial systems, *i.e.* systems/consortia which are not unique and change constantly. Therefore, some firms prefer to use their "inventions" and rely upon their expertise and know-how instead of making them public by patenting.

6. Attitude towards environmental biotechnology

In general, the developers and suppliers of environmental biotechnology products and/or services are quite realistic about the contribution biotechnology can make to management strategies for environmental problems. All of the companies are very well aware of the advantages and disadvantages of biotechnology compared to other environmental techniques. Most companies consider biotechnology as a technology for specific environmental problems where other techniques are not (cost) effective. However, there are also parts of the market where biotechnology competes directly with alternative techniques. Then, the degree of "belief" in biotechnological solutions is dependent on the background of the decision makers. Nevertheless a general trend towards an integrated application of biotechnology together with other environmental technologies is already visible.

It was found that biological methods for pollution treatment are perceived by the user sectors as more difficult to implement, because the polluting companies do not have biological expertise. Moreover, in the current state of environmental biotechnology, several processes and procedures are not very flexible, for example, variations in pollution concentrations in effluents are only dealt with with difficulty. It is also a more delicate matter to control them by automatic regulation systems. Some experts believe that R&D efforts in the pollution treatment sector to date have been essentially restricted to the area of fundamental biotechnological engineering science rather than being set in a perspective of improved capacity to manage biological processes.

Thus, improvement of operational stability and predictability under practical circumstances is regarded as a key factor in a successful application of environmental biotechnology. This in turn requires demonstration facilities for recently developed environmental biotechnology processes in order to reach the status of "proven technology" and convince the decision makers in industry and government.

Chapter 10

Environment for Innovation

1. Industry views on support and perspectives

General aspects

Throughout the previous chapters, several factors influencing the opportunities for environmental biotechnology were identified by industry. Generally, these factors are comparable to those which apply for the total environmental industry. Figures 28 and 29 show the relative importance of several of these factors as indicated by the environment industry in the Netherlands. In many countries, the most important positive factors for further development of environmental technology were considered to be legislation and enforcement, followed by support for technological innovation and implementation. The most important negative factors were lack and unpredictability of scientifically justified environmental legislation followed by economic factors such as return on investment (of R&D and equipment expenditures), foreign competitors and market saturation. Except for the last two items, it is national and international legislation which largely determine the development and innovation of environmental technology.

For the consuming industry (the users of environmental technology) a similar role is played by governments. Main issues here are:

Figure 28.   **Positive factors for the environmental market as perceived by the Dutch suppliers of environmental technology**

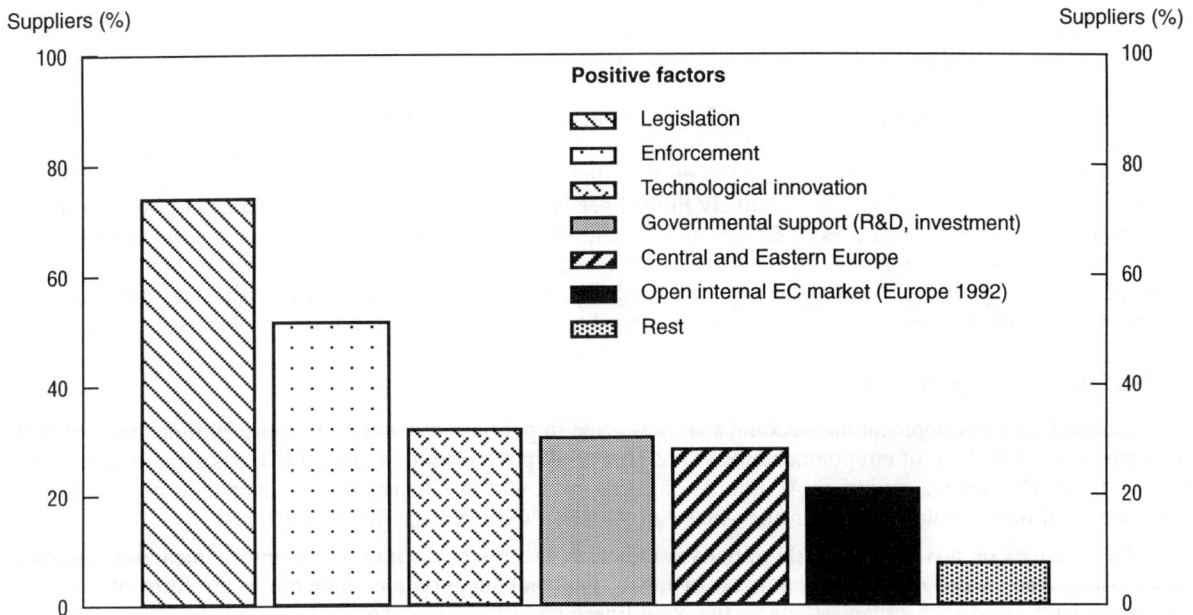

Source: VLM, 1991.

167

Figure 29. **Negative factors for the environmental market as perceived by the Dutch suppliers of environmental technology**

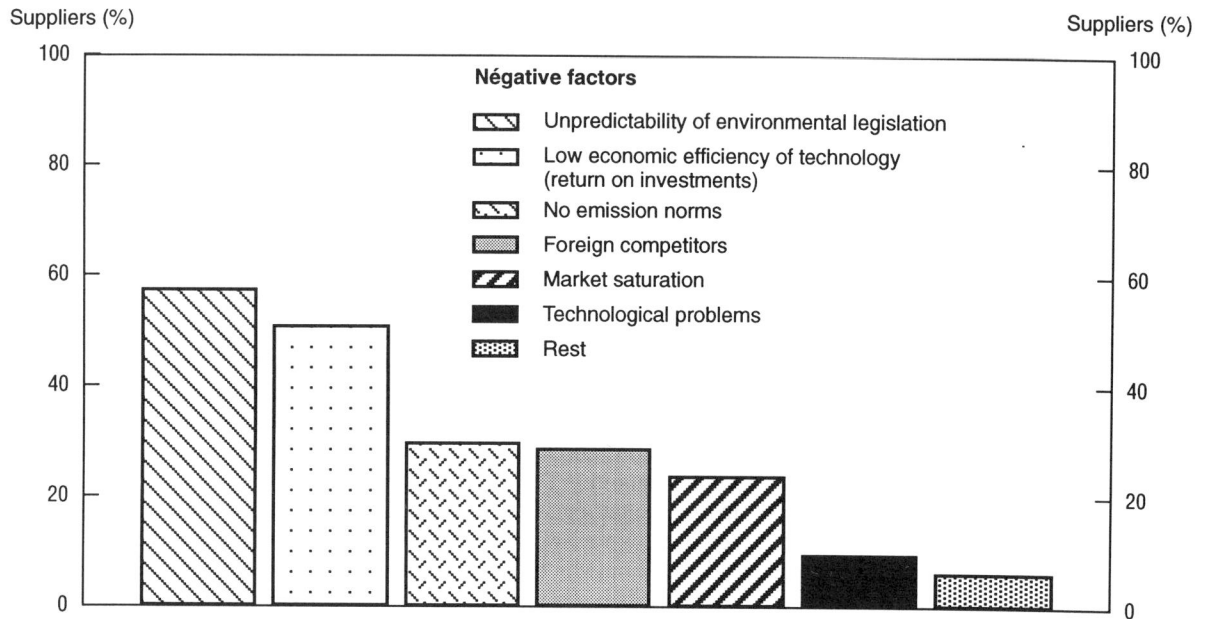

Suppliers (%)

Suppliers (%)

Négative factors

⬚⬚	Unpredictability of environmental legislation
⬚	Low economic efficiency of technology (return on investments)
⬚⬚	No emission norms
▨	Foreign competitors
▨	Market saturation
■	Technological problems
▦	Rest

Source: VLM, 1991.

- internationally harmonised legislation (as environmental costs are an important competitive factor) whereby the geographic and climatic differences between countries must be taken into account;
- standards and legislation which do not change unexpectedly and dramatically over time;
- setting of minimum requirements and freedom to meet these requirements in their own way (*i.e.* not making the latest technological development the new standard);
- application of BATNEEC principle (best available technology not entailing excessive cost);
- and the realisation that sometimes it is more prudent to clean-up 80 per cent of a problem for say, 20 per cent of the cost, rather than pursuing a 100 per cent solution at an extremely high cost.

It is understood that the same factors are at least as important for development and innovation of environmental biotechnology. Although there is considerable interest among Japanese industries in the use of biotechnologies in pollution control, most companies take the position that it is still too early to consider seriously the option because of a lack of seed technology, relatively limited needs compared to the high development cost, and severe restrictions on field trials. Industries in Europe and the United States take similar positions but seem to be slightly more optimistic on these matters. Nevertheless, as biotechnology is a relative newcomer in the environment industry, adequate public sector support for an extended period is regarded as indispensable, in order to make environmental biotechnology the technology for sustainable development that industry considers it could be.

R&D support by the public sector

Research and development has become a serious issue in public sector support. Table 29 indicates that both the amount and the share of environmental R&D as part of all government-supported R&D is steadily increasing for most OECD Member countries. Additional data can be found in Chapter 6 "An economic perspective on environmental biotechnology", and in the Annex to Part II, "Government R&D initiatives".

As the annex on government R&D initiatives indicates, incentives for innovative environmental biotechnology exist in many OECD Member countries. In Japan, financial supports are available in the form of commissioned research from such organisations as the New Energy and Industrial Technology Development Organization (NEDO), the Research Institute of Innovative Technology for the Earth (RITE), Japan Bioindustry Association (JBA), and the Water Re-Use Promotion Center. Universities and national research laboratories also

Table 29. **Public sector support for environment research and development**

Country	1975		1980		1985	
	Million ECU	Share of all R&D (%)	Million ECU	Share of all R&D (%)	Million ECU	Share of all R&D (%)
Denmark	n.a.	n.a.	3.6	2.1	6.6	1.5
Germany	53.2	1.0	102.0	2.0	309.3	3.1
France	35.5	0.8	46.7	1.1	51.1	0.5
Italy	5.6	0.6	13.6	1.0	44.6	1.0
Netherlands	n.a.	n.a.	n.a.	n.a.	53.7	4.1
Spain	0.0	0.1	2.2	0.6	9.2	1.0
United Kingdom	25.8	0.5	30.9	0.7	99.6	1.1
EC	140.0*		230.0*		600.0*	
United States	190.2	0.9	171.7	0.8	259.5	0.5
Japan	50.8	1.5	60.3	1.6	n.a.	n.a.

* Author's estimate.
n.a.: Not available.
Source: OECD, 1992.

offer technological assistance to the industry. In the United States, the EPA initiated the SITE, the Bioremediation Research programme (as a part of the Biosystems Technology Development programme) and the Pollution Prevention Research Plan which have biotechnological components. In European Community Member countries, a variety of national programmes exist on biotechnology or environmental technology which have strong or weak environmental biotechnology components (amoung others Germany, United Kingdom, the Netherlands and Belgium). Additionally, the EC has initiated transnational research on relevant topics in programmes such as STEP, ENVIRONMENT and LIFE.

A comprehensive report (for the Dutch Government and the Working Party on Environmental Biotechnology in the Netherlands) on developments in public sector R&D programmes on clean, environmental and/or biotechnology processes and sciences also concluded that an environmental biotechnology component can quite often be identified. The total value of this R&D is limited although many programmes exist.

There is a rather strong industrial plea to make public sector R&D support more substantial and targeted at the bottlenecks in environmental biotechnology. Industry requires:

– Long-term fundamental programmes and projects to improve insights in underlying biological processes and develop new principles.
– Shorter-term demonstration projects to test and develop environmental biotechnology processes under practical conditions.
– More focus for projects on soil treatment and less focus on air and solid waste treatment. Although waste water treatment is relatively mature, it should also get attention as clean water is considered as a limiting raw material for life, health and production.
– Development of process integrated/clean biotechnology.
– Development of biotechnological added-value and recycling processes.
– Including heavy metals, dioxins and PCBs, halogenated hydrocarbons, solvents, PAH, oil spills, and pesticide residues among target compounds.

Companies are actively looking for support to innovate in the relatively new field of environmental biotechnology and to develop and demonstrate new processes.

In countries with government R&D programmes on environmental (bio)technologies, clean technologies, and even on cleaning specific particular natural compartments, most of the companies interviewed participate in such programmes. Exceptions are some of the Model I companies whose strategy is not to use governmental R&D money in order to stay independent. On a European level some of the interviewed companies (mostly Model II companies) participated in EC programmes (EUREKA and some of the R&D-programmes of the third Framework Programme).

It is remarkable that none of the companies felt that national or international government support was indispensable for their R&D or demonstration activities in this field. With the exception of one small, one-person company, they could have done without. However, without this support, progress would have been made at a much lower rate and the type of approach would have been different: based on trial and error with black boxes rather than on insight and understanding. Fewer and less reliable products and processes would have entered the market or would enter it in the future.

The companies, asked whether it cost them any trouble to obtain (venture) capital, answered that it was rather easy to do so at present.

Human resources

The European labour market in the environmental sector is rather tight, especially in north-west Europe and particularly for technically qualified individuals. However, attracting young graduates with special knowledge of environmental biotechnology is no real problem for most of the interviewed companies. Some firms have a policy of recruiting junior employees for whom proper training will be provided.

Public attitude and education

Public concern about environmental issues is high and still increasing in Europe (Figure 30). Most of the issues raised can be positively affected by environmental biotechnology. In particular, the option of restoring

Figure 30. **Public opinion: local, national and world environment**

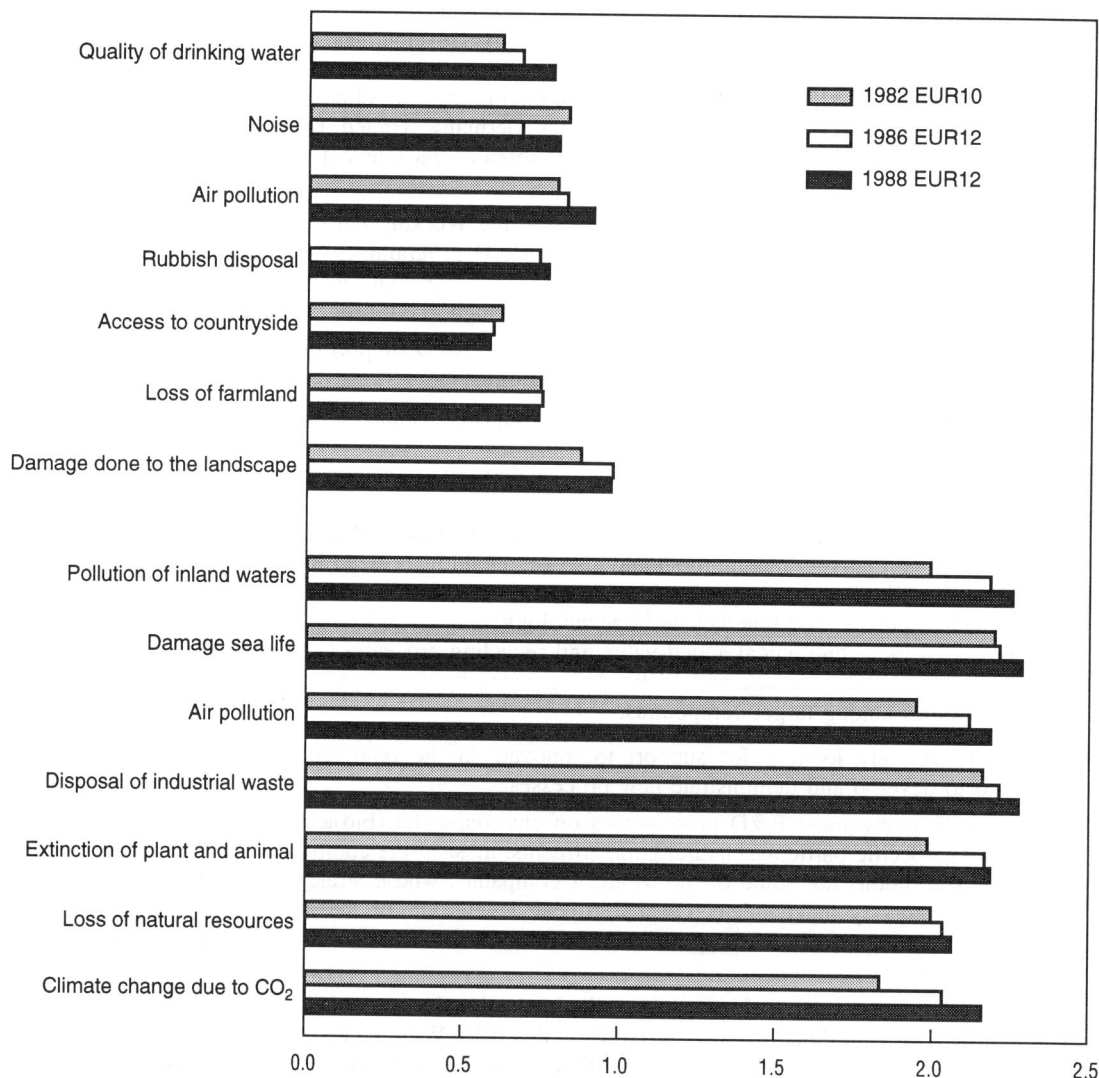

Notes: Values are mean scores for replies weighted with respect to their importance as follows: "a great deal" (3); "a fair amount" (2); "not very much" (1); "not at all" (0).
Source: "Opinions of Europeans on Biotechnology in 1991", *Eurobarometer 35.1*, Commission of the European Communities, 1991.

natural balances and cycles by biological means is greatly appreciated by the public. There is less debate and controversy therefore regarding the potential benefits of biotechnology for environmental purposes than there is, for example, for food and feed applications.

In a number of north-west European countries, the word "biotechnology" is almost synonymous with "rDNA technology". European industry would prefer it made clear that current research in environmental biotechnology does not actively involve rDNA technology. In these countries some social groups resist the use of these techniques in certain application areas (food and agriculture), but the "bad" name of biotechnology has

Figure 31. **Types of biotechnological research that are worthwhile and should be encouraged according to the public in the EC Member countries in 1991 and 1993**

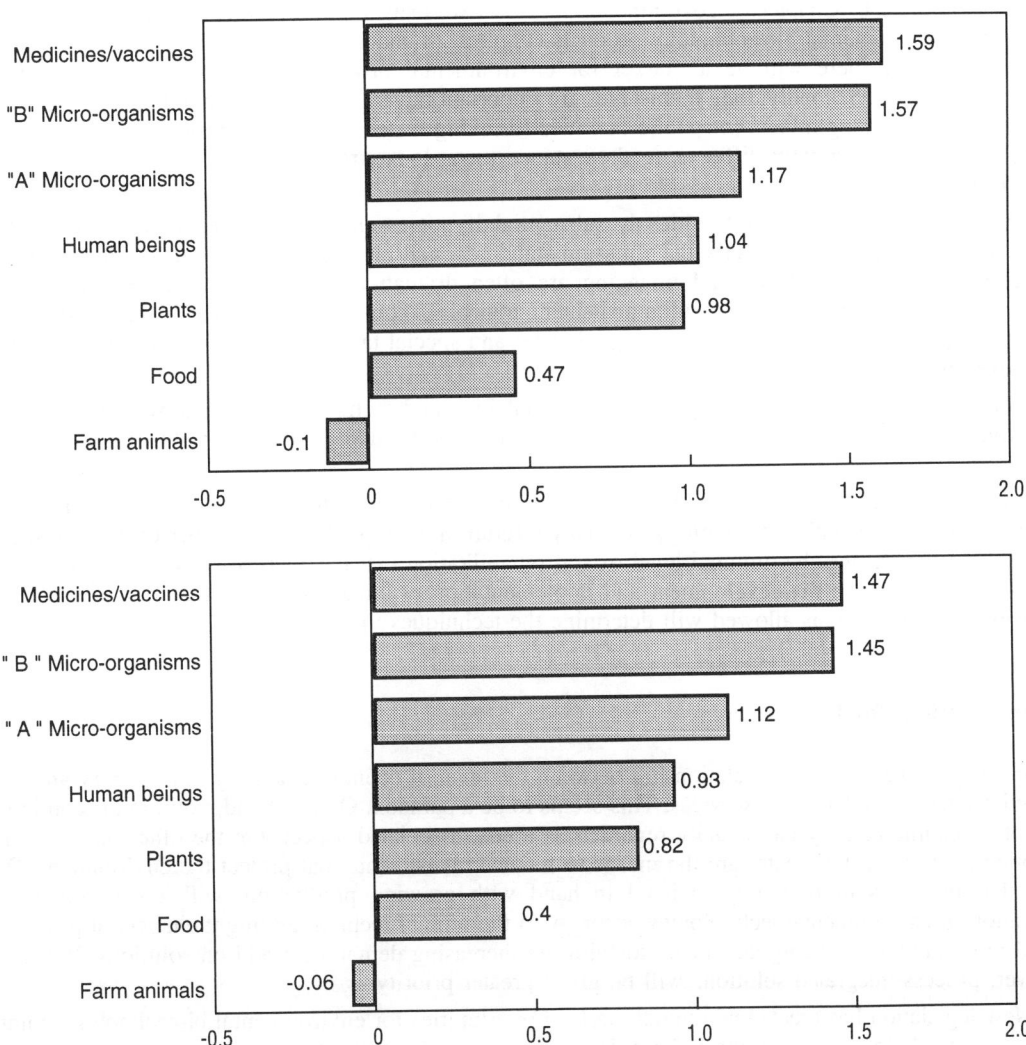

Notes: "A" micro-organisms are those such as yeast to make bread, beer or yoghurt; "B" micro-organisms are those used to break down sewage and other waste products in order to turn them into materials harmless to the soil. Figures are overall averages for the EC adult population: +2 = maximal support and -2 = minimal support.

Source: "Biotechnology and Genetic Engineering: What Europeans Think About It in 1993", *Eurobarometer 39.1*, Commission of the European Community, 1993.

diffused throughout the media. Public attitude is mentioned as one of the most important competitive disadvantages of biotechnology compared to other, traditional technologies applied in the environmental sector.

Although the public does not generally interfere in decision-making processes concerning clean-up activities, they do when official procedures have to be followed in order to receive permission for the use of GMOs on new locations. Furthermore, there is strong opposition to the deliberate release of engineered micro-organisms for special purposes. Thus, the focus of companies has shifted from the potential advantages of GMOs to operational stability, yields, efficiencies and cost-effectiveness of environmental processes using naturally occurring micro-organisms.

A more recent survey in the Member countries of the EC, specifically devoted to biotechnology and genetic engineering, confirms public support for environmental biotechnology at a level equal to medicines and vaccines (Figure 31). Up to 87 per cent agree or strongly agree with biotechnology research for environmental purposes. But nonetheless people believe that a regulatory framework is needed.

2. Role of government authorities and legislation

As mentioned in Chapter 6, governments play an extremely important role in directing and shaping the demand for environmental biotechnology products (Figures 28 and 29; Table 29). Firstly, governments strongly determine whether there will be a market for environmental products at all by means of legislation and compliance policies. Secondly, they themselves are important buyers of environmental biotechnology products in order to provide, among others, for public facilities (drinking water, sewerage) or clean land. As was explained above, countries may greatly differ as to their stance towards environmental pollution and the role of public policy in the management of it.

The role of government authorities is substantial if not essential since they dictate the need to install pollution control equipment, specify the emission standards, and monitor enforcement. In other words, demand, product performance standards, and financing are often dramatically affected by government action which thereby affects the economics of pollution abatement control. In some countries, financin for pollution abatement hardware is provided by the government (*e.g.* France) and special tax-based pollution abatement strategies have been developed.

Companies and innovation in the environmental bio- and technology segment will benefit from a firm legislative framework. This is also required to clarify safety issues and grant patent/industrial property protection rights.

Lasting political debates on whether priority should be given to, for instance, incineration or composting may put an environmental biotechnology company's future into jeopardy. In a number of countries, discussions are going on concerning the minimal level of soil rehabilitation that has to be observed. The outcome of these discussions is crucial to the development and implementation of biotechnological methods of soil treatment, as the residual concentrations allowed will determine the techniques to be used.

3. Implications for industry

In Europe, there is a clear relationship between the level of economic activity in a country and the demand for environmental products and services. This seems to be a paradox. On one hand, pollution is contingent on the level of economic activity and usually regarded as a cost-associated aspect. On the other hand, the higher the economic activity is, the greater are the means to finance investments that protect the environment. Only in the event that increases in pollution go hand in hand with growing production, will a substantial rise in the application of environmental technologies occur. Also the type of technology might change: at present, environmental biotechnology is being developed to fulfill an increasing demand for add-on solutions. In the long term, however, process-integrated solutions will be given greater priority.

New legislation has created additional market opportunities for environmental biotechnology companies. In the future, legislation will continue to be a decisive factor and greatly determine the opportunities for environmental biotechnology. There is a clear causal link between long-established, stringent legislation and the competitive strength, particularly in the export area, of the environmental industry. Good examples of this may be seen in the industries of Denmark, Germany and the Netherlands. Financial support for innovative R&D, as well as for demonstrations of recently developed processes, will be needed to influence the maturing of environmental biotechnology. This greatly determines the time frame in which environmental biotechnology will become the expected key technology in sustainable development.

Chapter 11

Conclusions

Biotechnology already makes an important contribution to pollution prevention, preservation and protection of the environment. Environmental biotechnology products and services are being developed and used for the treatment of pollutants at the source (add-on technologies) and the treatment of diffused pollutants (remediation). Presently the products concern relatively simple installations with natural micro-organisms as the active component.

"Classical" biological processes have already been applied for decades and account for about 15 to 25 per cent (about $40 billion world-wide) of the environmental market. The contribution of recent biotechnology developments to the overall environmental market is as yet rather small: it is currently valued at a maximum of 5 per cent of the total market. On the basis of an estimated total environmental market in Europe of about $300 billion in the year 2000, the biotreatment market can thus be valued at more than $50 billion. Overall, there is a lot of optimism for the future share of biotechnology in the environmental market. Environmental applications of biotechnology will produce large savings in the costs of pollution treatment.

Process-integrated biotechnology, sometimes used as a synonym for "clean technology" is still in its infancy. Nevertheless, it is generally accepted that biotechnology will increasingly replace chemical processes with enzymatic reactions. For most of the big multinational companies in this field, sustainable development is generally accepted and adopted. These companies have environmental policy statements which are more than just lip service. Thus, the development of clean production processes has become more and more an inherent part of the procedures and developments followed by these companies.

Diffusion of such a concept to small and medium-sized companies will have to be one of the focal points for policies for environmental technology. Many of these companies are unaware of the sustainability concept of clean technology.

In the long term, prevention of pollution will be a realistic strategy for managing the process of sustainable development. In the short and medium term, the use of add-on and remediation technologies will be indispensable. Nevertheless, in practice, all three management strategies will be followed. In a number of cases, add-on technologies are more simple and cheaper than process-integrated technologies. However, comparisons are difficult as the costs for these technologies will be internalised and are not (yet) visible for reviews such as this.

Biotechnology will not be the sole solution but is to be considered as one of the tools for total environmental management. The main applications are expected to be a combination of biotechnological with other environmental techniques (physico-chemical, thermal and mechanical). There will be a tendency to offer integrated systems to the market, containing a number of co-ordinated techniques in combination with service contracts and often as turnkey processes.

In Europe and North America, markets for environmental biotechnology products are essentially determined by the "severity" of regulations for environmental protection. Overall, the role of government authorities as a driving force in this sector must be stressed, as these largely determine demand by setting the regulatory framework. These authorities specify standards and hence largely orient the technological methods of environmental control. They contribute to the financing of investments in pollution abatement/clean up equipment and services. Although it is a legislation-driven market, recession has its influence since the authorities are not strict in applying the rules if the continuity of pollution-causing companies is in danger.

The further development of the environmental biotechnology market will be determined by considerations of a technical and economic nature, but also by the development of priorities given to ecological problems and hence the development of policies on combined economic development and environmental protection implemented at a supranational level.

Although biotechnology has a number of advantages over other techniques, the diffusion of biotechnology in the environmental sector is relatively slow. The advantages of improved processes and superior product specifications compared to traditional methods are often not valued or not known to the potential users. One of the reasons is that the engineers who have the responsibility for solving environmental problems tend to look for solutions inside their own areas of expertise. The chemical engineer deals with pollution by chemical procedures, the mechanical engineer deals with them through incineration, etc.

The biological and ecological knowledge of most engineers in the pollution treatment industry remains inadequate. Pollution treatment processes are "locked in" to the culture of dealing with problems from one's own background. The barriers to the application of biotechnological procedures for the treatment of environmental problems in non-biotechnology based companies must be broken down by the use of publications, demonstration projects and the introduction of biotechnological principles in basic engineering education (Box 3).

Box 3.   Summary of industry views and perspectives on environmental biotechnology

- The environment market and industry is determined by governments. There is a strong requirement for:
  - international priority setting and action plans;
  - consistent and harmonised legislation;
  - proper enforcement;
  - no dramatic and unexpected changes in legislation, *i.e.* time for industry to meet the standards and develop appropriate technology;
  - support by implementation of environmental (bio)technology.

- Environmental biotechnology is regarded as very promising and as a key technology in sustainable development.

- Both process-integrated and add-on and clean-up (bio)technologies are required. The share of process-integrated technology is expected to increase in the longer term.

- "Classical" biological processes are already a relatively important factor (15-25 per cent) in existing environmental technology (especially in waste water treatment) but the share of recently developed biotechniques is still small (< 5 per cent).

  An increase is expected but dependent on:
  - improved reliability and predictability of environmental biotechnology processes;
  - R&D, both for underlying scientific disciplines as well as for scaling up to practical conditions;
  - demonstration projects to develop and reach the status of "proven technology";
  - public sector support for R&D, demonstration projects, implementation of environmental biotechnology processes, education, public attitude and awareness.

As a concluding remark, it should be stated that industry regards environmental biotechnology as one of the key technologies for sustainable development. There is substantial optimism about the role environmental biotechnology will play in the near- and medium-term future provided the above-mentioned technical, legislative and financial bottlenecks are addressed efficiently.

Chapter 12

Bibliography

BEWLEY, R.J.F. (1992), *Bioremediation of Contaminated Ground*, Dames and Moore International, Manchester, United Kingdom.

BMFT (1992), *Biosensorik*, Bonn, Germany.

BRAUER (1991), *An Industrial Perspective on Environmental Benefits of Biotechnology and on Legal Controls*, Proceedings of the International Symposium on Environmental Biotechnology, Part 1, Ostend, Belgium, 22-25 April.

BURRILL, G.S. and K.B. LEE (1990), *Biotech 91: A Changing Environment*, Ernst & Young, San Franciso, California, United States.

COOMBS, J. and P.N. CAMPBELL (1991), *Biotechnology Worldwide*, COBIOTECH, CPL Press, Berkshire, United Kingdom.

Department of Commerce (1991), *Biotechnology in Western Europe*, International Trade Administration, Washington DC, United States.

DODDEMA, H.J. and B.A. HEIDE (1992), *Milieu-ontlastende Biotechnologie. Opinie-onderzoek onder industriën naar een mogelijke stimulering van biokatalyse in het kader van een IOP-programma*, TNO Department of Environmental Biotechnology, Delft, The Netherlands.

EC (1992), *The State of the Environment in the European Community,* Brussels, Belgium.

EC (1993), "Biotechnology and Genetic Engineering", *Eurobarometer 39.1*, Brussels, Belgium.

Ecotec Research and Consultancy Limited (1990), *Business Opportunities in Environmental Management in Eastern Europe*, NETT International Conference, 23 November.

Ecotec Research and Consultancy Limited, (1992), "The Development of Cleaner Technologies: a Strategic Overview", *Business Strategy and the Environment*, Vol. 1, Part 2, p. 51, European Research Press, Bradford, United Kingdom.

GIBSON, N.B., R.P.J. SWANELL, M. WOODFIELD, J.W. Van GROENESTIJN, L.B.M. van KESSEL, P.G.M. HESSELINK and V. HECHT (1993), *The Abatement of Volatile Organic Compounds (VOCs) by Biological Methods*, TNO Department of Environmental Biotechnology, Delft, The Netherlands.

GROENESTIJN, J.W. Van, and P.G.M. HESSELINK (1994), "Biotechniques for Air Pollution Control", *Biodegradation*, Vol. 5(1), Kluver Academic Publishers, Dordrecht, The Netherlands.

HESSELINK, P.G.M. and G.J. ANNOKKEE (1992), *Verkenning milieubiotechnologisch onderzoeh in Nederland,* SENTER and TNO Department of Environmental Biotechnology, Delft, The Netherlands.

HESSELINK, P.G.M. and M.H. STOOP (1993), *Perspectieven en ontwikkelingen in de milieubiotechnologie*, TNO Department of Environmental Biotechnology, Delft, The Netherlands.

JOHNSTON, N. (1992), *Water: Resource and Opportunity,* Centre for Exploitation of Science and Technology (CEST), London, United Kingdom.

JOYCE, F.E. (1989), "Market Opportunities for Environmental Protection Technologies in Europe", IFEST Conference, 13 October.

LITTLE, A.D. (1989), *Biotechnology and Dutch Industry*, Report to the Ministry of Economic Affairs, EZ, The Hague, The Netherlands.

MASCINI, M. (1992), *Report on Biosensor Technology in Europe,* Florence, Italy.

Milieumarkt (1993), *Milieumarkt top 10*, Stam, The Hague, The Netherlands.

Ministry of Economic Affairs (1990), *Biotechnology in The Netherlands: Vigorous and Varied*, EZ, The Hague, The Netherlands.

OECD (1992), *The OECD Environment Industry: Situation, Prospects and Government Policies*, Paris, France.

PRIEELS, A.M. (1991), *Development of an Environmental Bio-industry: European Perceptions and Prospects*, Report for the European Foundation for the Improvement of Living and Working Conditions Dublin, Ireland, 15 October.

RAUCON (1989), *Bioremediation of Hazardous and Industrial Wastes, Technology Assessment, Economic Evaluation and Market Potential, 1990-2000, A Multiclient Study*, November, Report No. R-FA9001, Raucon GmbH, Dielheim, Germany.

REISS, T. (1990), *Perspectiven der Biotechnologie*, Verlag TüV Rheinland, Cologne, Germany.

SCHIJNDEL, M.W. Van, and E.R. SOCZÓ (1993), *International Development in R&D Programmes on Environmental Biotechology*, Report 776101002, RIVM, Bilthoven, The Netherlands.

STOOP, M.H. and P.G.M. HESSELINK (1993), *De industriële toepassing van milieubiotechnologie in Nederland: inventarisatie van opties en knelpunten*, SENTER and TNO Department of Environmental Biotechnology, Delft, The Netherlands.

VLM (1991), *De milieuproduktiesector in Nederland*, Zoetermeer, The Netherlands.

Appendix 1

Issues for Discussion with Industrial Companies

The following issues were raised in a questionnaire submitted to, and discussed with industrial companies (see list of companies and persons interviewed in Appendix 2).

1. General economic aspects:

    a) Present and future markets; turnover and trends in market development.
    b) Origin of environmental biotechnology companies (spin-offs from larger companies, from university, from other biotech activities or conversion from other sectors).
    c) Attitude towards environmental biotechnology in big and small companies.
    d) Reasons for developing environmental biotechnology:
    –   site-specific or global environmental pollution and/or solution;
    –   present or expected pollution and/or sollution.

2. Company policy:

    e) R&D expenditure in environmental biotechnology in absolute terms and compared to other (total, environmental or biotechnological) R&D.
    f) Competitiveness/cost-effectiveness of environmental biotechnology compared to other environmental technologies.
    g) Role of university/industry/research institute network.
    h) Market heterogeneity; main company strategies (acquisition of technology, development in house, developments for market as a spin-off, developments for markets as core business).
    i) Patent position and strategy.

3. Differences in innovative environment (perception of companies):

    j) Public support for R&D/innovation (national and supranational)
    k) Effects (prohibitive or incentive) of environmental legislation on development of environmental biotechnology.

Appendix 2

Companies and Persons Interviewed

Biocatalysts Ltd, Pontypridd, United Kingdom: Tonney Godfrey, Director; Ian Johnson, Technical Sales Executive.

Bioclear Milieutechnologie BV, Groningen, The Netherlands: S. Keuning, Director.

Biodetox GmbH, Ahnsen, Germany: Dr. Axel Oberbremer.

Biotal Environmental Ltd, Cardiff, United Kingdom: Barry R. Twite, Director Biotreatment Ltd.

Biothane Systems Int. BV, Delft, The Netherlands: D.J. Koot, Technologist.

Clairtech BV, De Meern, The Netherlands: C. van Lith, Sales Manager.

Comprimo BV, Amsterdam The Netherlands: P.G. Paul, Senior Process Engineer.

Linde AG, Höllreigelskreuth bei München, Germany: Dr. Hans Reimann.

Mourik BV, Groot-Ammers, The Netherlands: Mr. van Dam, Head Soil Treatment and Environmental Techniques.

NOELL GmbH, Würzburg, Germany: Dr. Robert Klemps.

Pacques BV, Balk, The Netherlands: P. Hack, Director.

Rhône Poulenc, Direction Sécurité et Environment, Courbevoie and Lyon France: Lambert, Director of Research/ Environmental Expert; Lunel, Scientific Adviser, Biosciences, Scientific Directorate.

Shell Research Ltd. Sittingbourne Research Centre, England: John G. Purse, Section Manager Forestry; Chris S. John, Section Manager Treatment Processes; R. Watkinson, Section Manager Bioremediation.

SustainAbility Ltd, London, England: John Elkington, Director.

Tauw Infraconsult, Deventer, The Netherlands: L.G.C.M. Urlings, Head R&D.

Troy Environ, Environmental Technologies Holland, Hoofdorp, The Netherlands: J.D. Jungschläger, Sales Manager.

Vereninging van Leveranciers van Milieu-apparaten en-technieken, The Netherlands: Mr. Ubbens.

WS Atkins, Epsom, Surrey, United Kingdom: Jeniffer A. Kirtonne, WS Atkins Consultants Ltd, Science and Technology Division, Environmental Sciences; Barrie Mould, WS Atkins Planning and Management Consultants Ltd, Environment Division, Principal Environmental Technologies.

Umweltschutz Nord, Ganderkesee, Germany: Dr. Vollmer-Heuer.

Persons interviewed on specific issues:

Mr. L. Da Gama, Bio-industry Association, London.
Mr. J. Lunel, Chairman ORGANIBIO, and Rhône-Poulenc, Paris.
Mr. Maarten van der Kolk, Shell, London.
Mr. L. Abel, Eureco/Eureka, Denmark.
Holger Pedersen, Commission of the European Communities, DG XI-Environment.
Joanna Tachmintzis, Commission of the European Communities, DG XI-Environment.
Philippe Jean, Commission of the European Communities, DG III.

Participants in the SAGB/TNO meeting on "Biotechnology for a Clean Environment", 27 November 1992, Brussels:

B. Ager, SAGB
D. Brauer, Hoechst AG
C.M. Enzing, TNO
J.C. Evers, AKZO
P.G.M. Hesselink, TNO
J. Keddie, Smith Kline Beecham
S. Lambert, Rhône-Poulenc
V. Sanahuja, Sandoz
D. Schmetz, Monsanto Europe

The following persons have contributed to the drafting of "Industry Strategies and Perspectives" under Dr. Paul G.M. Hesselink:

Dr. C.M. Enzing
Dr. T.J.J.B. Wolters
Dr. H.J. Doddema
Dr. M.H. Stoop

Annex

Government R&D Initiatives

The following annex summarises government initiatives to support R&D in environmental biotechnology in ten OECD countries: Austria, Belgium, Germany, Italy, Japan, Netherlands, Spain, Sweden, Switzerland and the United Kingdom. The texts have been provided by the countries themselves, or have been drafted on the basis of material provided by them.

It is the dearth of comprehensive and internationally comparable statistics on environmental biotechnology R&D in general, and of government support for it in particular, which has led to this effort to gather at least some descriptive information. At present, few, if any, OECD countries have specific and comprehensive data on what they spend to support environmental biotechnology R&D, and no reliable international comparisons between national data are possible.

The reasons are threefold. Firstly, the definition of environmental biotechnology varies among countries, and even among agencies in the same country; more often than not, biotechnology cannot be neatly separated from other environmental technologies and related R&D expenditures. Secondly, the number of national agencies funding environmental biotechnology is larger than in any other sector of biotechnology which makes data-gathering more difficult. Thirdly, in addition to central or federal national agencies, numerous state, provincial and local authorities are involved in environmental technologies, and some of them are funding environmental biotechnology R&D. Virtually no OECD country has a data-gathering system which would include all these provincial/local data into national R&D statistics.

Of all OECD countries, the United States has probably the largest government R&D funding (federal) for environmental biotechnology, and (relatively) the best statistics.

As the definitions used by US federal agencies are variable, the funding data are not comparable in the strictest statistical sense. However, in spite of such limitations, these statistics (Table 30) show a pattern which could be relevant for other OECD countries as well.

The most revealing figures are at the end of the third and fourth columns. Only 2.1 per cent of all federally funded biotechnology R&D is spent on environmental biotechnology. However, spending for this particular field has been growing very rapidly in the last few years: it increased by 19.3 per cent for the years 1991-93. Also, the

Table 30. **R&D funding of environmental biotechnology by the US federal government, 1993**

Agency	Million $	Vertical	Environmental biotech. as % of total biotech. R&D	Environmental biotechnology growth in % 1991-93	Biotech. R&D as a % of total federal R&D (1992)
NSF	20.5	24.6	10.0	27.2	8.6
DOE (Energy)	18.5	22.2	7.6	58.9	28.9
EPA (Environment)	18.3	22.0	100.0	28.0	3.5
DOD (Defence)	11.5	13.8	13.3	8.8	0.2
USDA (Agriculture)	8.5	10.2	5.1	n.a.	15.0
DOI (Interior)	4.3	5.2	86.0	–7.5	0.9
DOC (Commerce)	0.6	0.7	4.6	–2.4	2.5
NASA (Space)	0.6	0.7	1.3	8.8	0.4
DHHS (Health)	0.5	0.6	0.0	17.4	32.0
Total	83.3	100.0	2.1	19.3	5.7

Source: Biotechnology for the Twenty-first Century, A Report by the FCCSET Committee on Life Sciences and Health, February 1992, FY 1993.

funding comes from nine federal departments and agencies – a large number – with the lion's share provided by the NSF, DOE (Energy) and EPA (Environment), which together cover more than two-thirds of the total.

Figures from other OECD countries indicate that the proportion of total biotechnology R&D devoted to environmental protection is small in most countries, ranging approximately, according to country and definitions, between 2 per cent and 10 per cent. Expenditures have been rising rapidly in the last few years, and reflect a growing number of initiatives from various agencies.

The salient fact remains the small percentage spent on environmental protection bears no relation to the perceived urgency of environmental problems in OECD countries or to the numerous specific research needs mentioned in the OECD workshop paper on "R&D Priorities and Policy Recommendations in Environmental Biotechnology".

The following description of government initiatives in ten OECD countries illustrates that initiatives and spending have been growing in accordance with the sense of urgency. What is also emerging is the decentralisation of initiatives in a number of countries, both in a geographical (national/local) and in a political sense (various departments).

There are not only similarities, but also important differences among these ten countries. Most noticeable is the difference between Japan's bold and forward-looking perspective which sees environmental biotechnology as a critical instrument to address the vast global problems of the 21st century, and the more pragmatic preoccupations of other countries, which focus on the immediate national and local problems of air, water and soil pollution.

1. Austria

Funds provided by the Austrian government for R&D in environmental biotechnology support a wide spectrum of basic and applied research and selected specific industrial applications, namely in the area of biological air emission reduction (biofilters, bioscrubbers), soil remediation and related chemical engineering aspects, as well as specific applications of waste water treatment. From 1985 to 1992, R&D in this area was supported within the framework of a targeted research programme on Biotechnology co-ordinated by the Federal Ministry for Science and Research.

The Federal Ministry for the Environment, Youth and Family is currently reviewing the Austrian R&D profile in environmental biotechnology in order to identify key areas for future targeting of its R&D policy and related funding activities. This should then provide the basis for an R&D programme to support the introduction and diffusion of innovative biotechnologies for the prevention, reduction and abatement of emissions and waste streams in various sectors of industry.

The Federal Environmental Agency is engaged in the evaluation of environmental biotechnology. A project under way will contribute to a comparison of different abatement technologies with special emphasis on biotechnologies including economic aspects. In addition it will describe "the state of the art" as well as the current diffusion of environmental biotechnologies in Austria. It will also investigate possible restrictions hindering increased application of such biotechnologies. The final report is intended to provide technical and economic information for possible users as well as policy recommendations including future funding needs.

2. Belgium

As a result of the successive stages of state reform, legal competence for environmental policy matters in Belgium has been almost completely decentralised from the federal level – which remains in charge of regulating product norms – towards the regional level. Moreover, as in other EEC member countries, environmental policy activities are increasingly determined by the growing framework of Community directives.

For reasons of coherence and synergy, the responsibilities for public support of environmental (biotechnological) R&D have also mainly been transferred to the same regional authority level. As in some other OECD countries, no specific environmental biotechnology programmes have been launched; R&D projects in this field of growing importance can therefore – in the first instance – be found within the ongoing biotechnology and/or environmental technology programmes.

In line with this overall organisational situation, the following sections give an indication of the state of affairs within the three Belgian Regions, preceded by a description of the role of the federal governmental level.

Federal Government

The Federal government remains in charge of the regulation of product norms and related research (investments in the field of environmental biotechnology research are negligible at present) and the organisation of the concertation – on local and international environmental and R&D-policy issues – among the different authorities concerned (State, Communities and Regions).

Moreover, in the framework of the government's fiscal policy, R&D-expenditures by industry directed towards environmental improvements qualify for increased tax deductions.

Brussels Capital Region

Linked with the particular characteristics of being a metropolitan region and its rather recent creation, some of the initiatives underpinning the environmental policy are still under development, and the policy yet to be implemented.

The "Brussels Institute for Environmental Management (BIEM/IBGE)" has a special section for "Clean technology". The latter's mission is to fund research and disseminate information relating to pollution prevention. No R&D-projects in the field of environmental biotechnology are financed at present.

Flanders

This region covering the Northern part of Belgium is coping at different levels with the growing environmental concerns. There has been an adaptation of the legal framework (VLAREM), as well as the set up of a selective investment support policy and – last but not least – an underpinning R&D-policy.

Within the last policy line, two main activities should be referred to in a priority way, namely: the "Flemish Environmental Technology Programme (VLIM)" and the "Flemish Action Program on Biotechnology (VLAB)". Both programmes are managed by the Flemish Institute for the Promotion of Scientific-Technological Research in Industry (IWT).

VLIM was initiated by the Flemish Ministry of Environment in order to stimulate companies and research institutes to go for R&D-projects generating new technologies for the improvement of the environment. It has a budget of approximately 600 million BEF (ca. 15 million ECU) and runs from 1990 to 1994.

Half of the 25 projects that have been selected up until the end of 1992 are directly concerned with environmental biotechnology. They represent a public investment of about 5 million ECU (= ca. 50 per cent of the already allocated VLIM budget) and are distributed as follows: water (30 per cent), air (10 per cent), waste (20 per cent) and soil pollution (40 per cent); or: prevention (negligible), detection (10 per cent) and remediation (90 per cent).

The key issues are elimination of nitrogen and phosphorus from waste water, elimination of organic micropollutants such as EDTA from water, purification of air by microbial catalysis, odour problems in anaerobic water purification, biological treatment of soils contaminated with mineral oil, PCBs and PAHs, anaerobic composting and sludge treatment.

VLAB is an initiative of the Flemish Ministry of Economic Affairs to stimulate R&D initiatives in specific subfields of modern biotechnology. The programme has a global budget of about BF 900 million (ca. ECU 24 million) and covers a five-year period (1990-94). Recently, another BF 620 million (ca. ECU 15.5 million) has been allocated to support strategic research in Centres of Emerging Technology within the next five years (1993-98).

Some of the precompetitive industry-university projects (max. 50 per cent financing) in the field of industrial microbiology and plant biotechnology clearly contain an environmental component and might, as such, contribute to more sustainable methods of industrial and agricultural production (30 per cent of a budget of ECU 11 million).

Within the area of more basic research with strategic industrial potential, ECU 0.5 million (6 per cent of the total budget) has been allocated to the Flemish Institute for Technology Research (VITO) in order to study, in selected micro-organisms, the molecular mechanisms of heavy metal resistance and breakdown of chlorinated biphenyls.

Moreover, within the accompanying portfolio of technology assessment projects, one deals with the release of GMOs.

There is no doubt that, apart from these two major schemes, some remaining environmental biotechnology R&D projects are funded by other administrations, such as the Ministry of Education, or by semi-public organisms, such as VLAR and AQUAFIN, dealing respectively with waste and water management.

The VLAREM legislation includes the obligation to construct a plant according to ''the best available technology not entailing excessive costs'' (BATNEEC principle) and specifies environmental quality objectives and emission standards.

Public investment support is selectively tied to the preventive environmental performance of the proposed production plant, which is determined by an environmental impact study.

Wallonia

Science and technology policy is a major element of the efforts to transform and promote innovation in Wallonia's economic activities. Biotechnology, as well as the development of clean technologies, clearly represent priorities within this overall policy line.

The ''Directorate-General for Natural Resources and the Environment'' is in charge of the preservation of Nature and Forests and of the management of water, subsoils and wastes. However it is another administration, ''Directorate-General for Technology, Research and Energy'', which stimulates R&D involving industries, research institutes and universities through different programmes (First programmes, R&D projects, R.I.T.). Some of those projects concern environmental biotechnology (biological purification of industrial effluents and gaseous emissions). This represents a public investment of BF 200 million (ca. ECU 5 million).

3. Germany

Germany has extensive experience with biotechnology and the environmental applications of biotechnology (particularly the treatment of waste water). This is also a measure of the government's response to growing public concern over social and political issues, particularly with regard to the consequences of reunification and the environmental legacy of the former East Germany, as well as the emerging political influence of the ''green'' parties.

In Germany, as in some other countries, there is only a limited amount of direct State aid for environmental biotechnology. Most of the legislation and regulatory measures in place apply to biotechnology in general (such as the Genetic Engineering Act, which provides a legal framework for R&D relating to genetic engineering and its applications, and the Plant Protection Act which provides support for biotechnological innovation aimed at countering the adverse effects of the use of chemical biocides). Other measures concerning general protection of the environment (such as the Federal Emission Control Act which tackles the problem of air pollution, as well as the Waste Products Act) may use biotechnological means, bio-filters for example, to help prevent or reduce emission.

In addition to all these measures in support of biotechnology and the environment, the government also sponsors research into safety criteria and standards (for workers and public health) relating to the use of environmental biotechnology.

The funding programme announced by the Minister for Research and Technology in February 1992 indicates the scale of the government's commitment to the use of biotechnology to combat environmental problems in that it encompasses not only prevention (through the use of biodegradable materials, recycling of waste) but also treatment (purification, decontamination).

Applications for new technologies are actively promoted by setting up agencies to promote technology transfers from research institutes to industry. These are aimed at providing the information and training needed to apply these new technologies. They also have to raise the awareness of policy makers in local government and the private sector with regard to environmental biotechnology and related processes.

With regard to links between the public and private sectors, priority is given to co-operative projects between research institutes and private firms. Preference will be given to research projects or proposals put forward by technology transfer agencies which involve partnerships with institutions from the former East Germany.

4. Italy

The Italian government has not developed specific policy with regard to support for R&D relating to environmental biotechnology, particularly with regard to exchanges between the public and private sectors. The funding supplied for such research, however, has been relatively substantial.

In Italy, funding for R&D relating to environmental biotechnology is provided under the following four programmes:

  i) National Research Programmes (NRP), which include programmes on advanced biotechnology and environmental protection. The R&D activities relating to environmental biotechnology include:
  – biological treatment of liquid and solid agricultural wastes;
  – biological treatment (bioremediation) of contaminated soils;
  – development of biosensors for the monitoring of drinking water.

  ii) Mission-oriented projects (‘‘progetti finalizzati’’) set up in 1976 by the National Research Council, whose ‘‘Environment and Territory’’ project, which includes activities that might potentially involve biotechnologies, was presented to the Italian scientific community in early 1990, but as yet has not been implemented.

  iii) Waste disposal programmes run by the National Research Council. The Ministry of the Environment, with the approval of the Ministry for Scientific Research and Technology, is currently funding R&D programmes relating to the disposal of solid wastes, the treatment of liquid effluents, and the incineration of sewage sludge.

  iv) Three-year Plan for the Environment 1989-91.

This plan was set up under the overall environmental policy of the Italian government. Part of the programme is concerned with scientific research and the environment.

With regard to links and exchanges of information between the public and private sectors regarding R&D in the field of environmental biotechnology, the study of Western European countries made in 1987 by the US Department of Commerce highlighted the lack of co-operation between biotechnology researchers in the public and private sectors in Italy.

Government support for private firms working in collaboration with public laboratories is mainly provided through national research programmes.

5. Japan

The care of the global environment is becoming the most important problem to be solved in the world today. A radical reform of the current industrial technology system is in progress as we approach the 21st century. The current system is based on the technology revolution of the primary sector, around 10 000 B.C., that brought about expansion of agricultural production and civilisation and on the technology revolution of the secondary sector, starting in the 17th and 18th century, that led to the expansion of manufacturing industry.

The current industrial technology system is totally dependent on the irreversible consumption of fossil resources. It is not only causing problems such as the exhaustion of natural resources, but also accumulation of man-made and discharged substances that accompany the growth of population and expansion of human activities.

Japan believes the current industrial technology system should be transformed into a new industrial technology system for the coming 21st century.

This new industrial technology system must be in harmony with the global materials cycle in the biosphere. It should be gentle to humans and to the environment, and have four key characteristics: the renewability of resources; the mildness of production processes; the compatibility of products with humans and the environment; and the recyclability of wastes. We expect that biotechnology will play a major role in this transformation to a new system.

In the first category, CO_2 can be taken out of mere exhaust gas, and used as material to be converted biologically and chemically into useful substances such as fertiliser, feed-stuff, acetic acid, formic acid, and so on.

The second category branches into two parts. In the first part, we develop materials with low environmental burden, to substitute for substances that are difficult to decompose in the natural environment. In the second part, development of technologies for recycling and decomposing of utilised material is being undertaken.

In the third category, the aim is to convert the production technology itself so that it will be in harmony with the environment. For example, acrylamide in the past has been produced from acrylonitrile through a high-temperature, high-pressure, petro-chemical reaction process. However, as a result of a switchover in methods to the use of bioreactors, it can now be produced by biochemical reaction, with a conversion efficiency of almost 100 per cent, at normal temperature and normal pressure, simplifying the refining process by a large margin.

The Ministry of International Trade and Industry, MITI, has been evaluating policies for the future development of biotechnology.

Since biotechnology is expected to be a pathway to the 21st century, R&D must be considered of the utmost importance. It is hoped that the outcome of this R&D will make a broad contribution to the world by increasing human knowledge.

Currently, MITI is progressing with plans to establish the Research Institute of Innovative Technology for the Earth (RITE). The idea is to establish a research institute to promote basic research for the creation of innovative technologies that contribute to the improvement of such global environment problems as carbon dioxide which may cause the "greenhouse effect", man-made chemical substances which cause ocean and lake pollution, etc.

Biotechnology also possesses a huge potential in many areas still unknown. International co-operation is useful for the promotion of basic research and for maintaining co-ordinated safety measures.

Japan expects to contribute to global scientific and industrial progress through basic research and the promotion of exchange in biotechnology, and through technology co-operation with developing countries.

6. The Netherlands

The Netherlands has long been renowned for its expertise in the fields of microbiology, biochemistry and process engineering.

A recent study on "The Situation of Biotechnology in Europe" estimated that the Netherlands would become leader in five (the first five cited) of the following seven major fields of applications/products: health care and pharmaceutical products, agriculture, food and animal feedstuffs, chemicals, waste treatment, energy, and plant technology.

The initial decision to overhaul national science policy in the Netherlands in order to provide a better response to the economic and industrial needs of the country was taken in 1979. This was followed by the creation and gradual implementation of a series of Innovation-Oriented Research Programmes (IOPs), in which two ministries, the Ministry of Economic Affairs and the Ministry of Education and Science, were initially involved.

R&D programmes relating to environmental biotechnology were first developed under the IOP and other programmes.

Measures to protect the environment introduced under the Dutch National Environmental Programme (NMP) are aimed primarily at preventing polluting emissions at that source and at processing wastes in a way which does not generate non-recyclable end-products. To achieve this goal, the government has directed its main efforts towards providing incentives for R&D carried out under various research programmes as well as funding for research programmes, the main ones being as follows:

i) Innovation-Oriented Research Programme for Environmental Technology (IOP-MT). R&D relating to environmental biotechnology (IOP-MBT) is one of three sub-programmes in the above programme, the other two being recycling and prevention.

ii) Regulation for the advancement of environmental technology (stir MT), a funding programme aimed at the promotion of environmental technologies (under which support has been provided for R&D in environmental biotechnology).

Whereas the IOP-MT programme is aimed solely at universities and research institutes, the stir MT programme is also open to firms and gives priority to the prevention and treatment of persistant organic compounds (nitrogen and phosphorus) in waste water, polluted soil, waste air and solid waste (including heavy metals).

In addition to the two programmes mentioned above, there are several other programmes which are much broader in scope, but under which some funding may be made available to R&D projects relating to environmental biotechnology:

i) First, a programme providing subsidies for industrial research projects (PBTS) relating to biotechnology (PBTS-BIO) and environmental technology (PBTS-MT). Any research project relating to environmental biotechnology may therefore be considered under one or the other of these programmes.

ii) The integrated soil research programme. A small part of this programme (10-20 per cent) is devoted to the biodegradability of contaminants and the development of biological techniques for treating contaminated soil.

iii) The NOH programme, including the recycling of the organic fraction of domestic solid waste and treatment of manure.

iv) The RWZI-2000 programme, including the biotechnological treatment of municipal waste water, sewage sludge and treatment of manure.

v) The POSW programme on treatment of sediments.

A number of Acts and Decrees also provide indirect support for R&D in the area of environmental biotechnology. For example:

– The Interim Act on Soil Contamination.

– The Nuisance Act and Air Pollution Act, which have helped promote the use, in particular, of biofilters. Since 1978, over 200 such filters have been installed in the Netherlands.

– The Surface Water Pollution Act. As a result of this Act, all waste water in the Netherlands now receives some form of biological treatment.

– The development of non-chemical products to treat soil and plant diseases.

– The Decree regulating the use of genetically modified organisms; part of the research in the area may be considered to relate to environmental biotechnology.

The basic framework for links and exchanges of information between the public and private sectors with regard to R&D in the field of environmental biotechnology is provided by the IOP-MBT programme (Innovation-Oriented Research Programme relating to Environmental Biotechnology).

7. Spain

Biotechnology and environmental R&D are grant-supported in Spain by the central government, but also by the autonomous governments of the 17 autonomous communities. This summary refers only to the central government efforts.

The Spanish Law for the Promotion and General Co-ordination of Scientific and Technical Research, approved in 1986, established the National Plan for Scientific Research and Technological Development (National R&D Plan). This plan has set up priorities, initially for the period 1988-91, and later on for 1992-95.

Specific programmes were created under the National R&D Plan. The National Programme of Biotechnology (NPB), started in 1988, represents the continuation of the Mobilisation Programme in Biotechnology of 1986-87. It has focused initially its interest in five priority areas: *i)* basic biotechnology research; *ii)* agriculture and food; *iii)* human and animal health; *iv)* industrial processes; *v)* biodegradation and pollution control.

During 1988-91 the financial support of the NPB has been close to Ptas 7 billion. The programme has grant-supported a total of 160 research projects, 22 of them corresponding to the area of biodegradation and control of pollution. A total of 97 full-time researchers work on these projects.

During the same period of time, 29 concerted projects between industries and research centres have been approved; only one related to environmental applications of biotechnology.

The National Programme of Conservation of the Natural Patrimony started in 1989 and has evolved in 1992 to an Environment Programme by fusion with parts of other programmes (*i.e.* Forestry, Aquaculture and Marine Resources, Geological Resources). The priority objectives in this programme for the period 1989-91 have been: *i)* desertification; *ii)* pollution; *iii)* continental aquatic systems; *iv)* marine systems.

The budget of the programme for the three-year period has been Ptas 2.2 billion.

One hundred research projects have been funded in which 338 full-time researchers are working. Eight of these projects are in the field of biotechnological applications. Thirty full-time researchers have been grant-supported in these eight projects.

Eighteen concerted projects have also been funded.

Biotechnology and environmental R&D are also funded in other National Programmes, such as agriculture, husbandry, health, fine chemistry etc., and biotechnological procedures to obtain foods, fine chemicals, transgenic crops resistant to different pests, etc., are fully supported through these programmes.

A number of government policies and laws also encourage R&D in environmental biotechnologies by indirect means, particularly the Law for Protection of the Atmospheric Environment (1972), the Law of Water (1985), the Law on Toxic and Dangerous Wastes (1986), the Law of Coasts, the EC Directive on the Treatment of Municipal Waste Water, and the National Plan on Industrial Wastes.

Transfer policies between public and private R&D include the joint project between public research centres and enterprises in the National Plan for Scientific Research and Technological Development. There are other actions, such as the Programme for the Transfer of Research Results (PETRI). This programme stimulates the link between public research groups and the private enterprise by financing specific projects for which a private (or public) enterprise/industry has made a declaration of interest.

The Ministry of Industry has specific programmes directed to stimulate R&D in enterprises.

- The PATI (Plan of Technological and Industrial Stimulation) offers grants for R&D for a three-year period in enterprises (budget = Ptas 6.8 billion). This programme a subprogramme on biotechnology, fine chemicals and materials.
- A specific programme on the environment (PIDMA, Environmental Industrial and Technological Programme) has been launched with the aim of modernising the Spanish industry and adapting it to the new environmental legal framework. The programme finances:
  - development projects;
  - demonstration projects;
  - application of non-pollutant new technologies to productive processes;
  - equipment development;
  - contract with enterprises in the environment sector to develop diagnostics, risks assessment, etc.

8. Sweden

The Swedish environmental protection agency and some other research councils such as the National Board for Industrial and Technical Development give priority to research on biotechnology for a sustainable development.

Biotechnology can support environmentally sound development directly, by environmental biotechnology, or indirectly, by the choice of raw materials or processes in industry, agriculture, forestry or aquaculture.

In a report to the Ministry of Environment and Natural Resources, the research councils state that biotechnology can be the best available environmental technology to prevent and solve environmental problems in water treatment, waste reduction, and bioremediation. Traditionally, biotechnology is used in environmental protection, but the councils foresee several applications in alternative agriculture, energy production, and chemical substitution.

Depending on the financial outcome of the proposal to the ministry, further planning will result in a prioritised research programme.

There is no specific law covering biotechnology/gene technology in Sweden. A parliamentary commission will later this year (1992) propose policies, jurisdiction and organisation of biotechnology risk control, especially in connection with genetically modified organisms.

Two research organisations are actively linking public and private environmental biotechnological R&D; Institute of Air and Water Research and, to some degree, the Biotechnology Research Foundation, Lund.

Nordic Fund for Technology and Industrial Development is supporting biotechnological R&D.

9. Switzerland

Two basic facts have shaped Swiss government policy for environmental protection.

Since natural deposits of minerals are scarce in Switzerland there is no mining industry. The last aluminium plant is about to shut down, asbestos production is almost at a standstill, and heavy metal engineering, which is confined to a few firms, is progressively diminishing. Indigenous energy sources (hydropower) cover barely 20 per cent of the country's requirements, and only 38 per cent of total energy is produced without CO_2 emission (hydropower, nuclear energy and wind power). Farming is affected both by competition from other countries and by the GATT agreements, while the traditional farm subsidies (hill farmers in particular receive major subsidies) are gradually being replaced by direct payments. In the lowlands, small farms are being ousted by large rationalised agro-industrial enterprises. As a result of all these trends, there is a high population density in Switzerland, farming is intensive, and commercial and industrial enterprises are modern.

The second basic fact is the continuing extensive autonomy of cantons and communes, which restricts the freedom of federal action in many environment-related fields (*e.g.* road system separate from the national highway network, erection of "téléphériques" and ski lifts, building legislation). Cantons and communes often fund R&D projects from the operational budgets for their installations. Because communal, cantonal and federal activities overlap in this way it is difficult to assess the overall research and development effort in environmental biotechnology in Switzerland.

The Confederation's environmental protection regulations are aimed at minimising air, water and soil pollution. Federal legislation covers all emission sources (including noise, for instance) as well as landscape preservation. The technical directives in this field must be complied with by the cantons and communes in order to obtain federal subsidies for their installations. In fact this system gives the Confederation an indirect but effective means of exerting pressure. In the Confederation's overall budget the "environment" budget increased from SF 11 127 million in 1988 (0.13 per cent of the budget) to SF 39 533 million in 1990 (0.39 per cent of the total), a 355 per cent increase. Furthermore, in 1992 the Federal Council launched a priority programme for environmental research with an overall budget of SF 35 million for the first four years.

Under these measures, biotechnology research focuses on waste water treatment and effluent treatment for agricultural purposes and, to a lesser extent, the production of biogas by means of combined aerobic/anaerobic installations. Switzerland produces 3 million tonnes of organic waste, about one-third of which is used for biogas production.

Effluent treatment using anaerobic-thermophilic bacteria is booming. This process has the two-fold advantage of being clean and of producing stable granular residues. Efficient biological processes are being studied for removing recalcitrant toxic compounds from the liquid effluents of manufacturing plants.

Solid waste is composted, incinerated, recycled or dumped. As far as possible, federal, cantonal and communal authorities encourage separate collection and recycling of metals, glass, waste paper, and plastics.

Biological processes can also be used for decontaminating soil and ground water polluted by industrial activities although this problem has seldom affected Switzerland up to now. So far, biological decontamination methods have been developed on an empirical basis and there are few well-established processes that are widely accepted. The subjects of controversy chiefly concern the preparation and maintenance of active micro-organism populations.

It is vital to improve the effectiveness and profitability of biotechnological processes in the future. For the time being, such developments are limited by the lack of suitable measurement methods and the absence of technical and economic optimisation of natural material cycles. Finally, the biological and technical basis for the disposal and recycling of recalcitrant waste and toxic compounds still needs to be established.

10. United Kingdom

In 1992, the UK government established the Office of Science and Technology (OST) to oversee science and technology policy. This was followed by a nation-wide consultation and review of both policy and organisation, the results of which were published in a White Paper in 1993.

The strategy which emerged is intended to improve the nation's competitiveness and quality of life by maintaining the excellence of science, engineering, and technology by:

- developing stronger partnerships between the research base and industry;
- supporting the advancement of knowledge and education;
- contributing to international and particularly European research effort;
- promoting public understanding;
- ensuring the effectiveness of government-funded research.

Mechanisms will be put in place so that:

- industry and the science base will be drawn more closely into policy discussions on the size, direction, shape and content of future research programmes;
- closer contacts between the science base and industry will be promoted and, while not downgrading scientific excellence as a criterion for supporting research, placing greater emphasis on relevance (relevance meaning the extent to which outcomes could be taken up by potential users);
- a Technology Foresight Programme will encourage closer co-operation between managers and scientists and will aid the identification of those areas of technology of most importance to the economy and welfare of the country.

A number of government-sponsored schemes exist to promote collaborative R&D in the environmental area, some covering all aspects of environmental science and others purely for environmental biotechnology. LINK is the main government initiative, now under the auspices of the OST, for supporting collaborative R&D in all disciplines and presently consists of 32 programmes (government commitment of £215 million by the end of 1992), involving nearly 400 companies and 200 academic research groups. Two out of every five companies are SMEs. Nine programmes are in the biosciences and perhaps four of these have peripheral environmental connections. As of April 1993, a LINK programme devoted specifically to environmental biotechnology has been active. This supports focused research on water and soil bioremediation and will involve a range of disciplines from civil engineering to molecular biology and microbial ecology. The budget for this programme is £9.2 million over the next five years, half to be provided by government sources and half to be met by the collaborating industrial companies.

The UK Department of Trade and Industry (DTI) promotes collaboration between UK companies, UK and European companies, and also between UK companies and research institutes. It runs a number of financial support programmes which provide grants to projects, usually up to 50 per cent of the total cost with priority given to small and medium-sized enterprises (SMEs). Research institutes include not only universities and government laboratories but also polytechnics and colleges of further education. The DTI also supports a number of information exchange clubs that bring together academics and industrialists. In particular, the DTI has supported the formation of such a club as a forum for the development of R&D collaboration in industrial effluent, wastewater and soil biotreatment.

Other environmental programmes which are supported either solely by the DTI or in conjunction with other government departments are ETIS, DEMOS, and Euroenviron, all of which have been put in place over the last four years. Each of these programmes covers a very wide field but includes biotechnology. ETIS, which is a joint programme with the Department of the Environment, has a budget of £12 million. Its primary aims include improving environmental standards and developing cleaner technology, and also improving competitiveness among users and suppliers of technology.

DEMOS has a budget of £4 million and supports the demonstration of plants and technologies in the environmental area with the intention of encouraging their wider use across the industrial sector. There had been approximately 500 applications (9 approved – one or two biotechnology) for support from DEMOS by mid-1992.

Euroenviron is a subgroup of the EUREKA programme which brings together companies from different European countries into R&D collaboration. It is concerned solely with environmental activities and will give support to environmental biotechnology. So far, Euroenviron, with a budget of some £11 million, has committed £4 million to 15 major collaborative environmental projects, none specifically in the environmental biotechnology area.

SMART and SPUR are schemes offering grants to small companies to develop novel ideas. The budget for SMART in 1991-92 was £12 million and in 1992-93 £13 million. Approximately 25 per cent of successful applications are for biotechnology. The SPUR programme has a three-year budget (starting 1991-92) of £30 million.

As in other countries, it is difficult to make firm estimates of government expenditures on R&D in environmental biotechnology. In part this is due to difficulties in accounting because of the lack of universally accepted definitions. Also, there may be double-counting because of the multi-disciplinary nature of the subject and because support for university and research institutes comes from different sources and can easily overlap. Best estimates are that biotechnology research represents some 4-5 percent of all civil R&D spending (UK civil R&D costs approx. £2.8 billion per year). Environmental biotechnology is approx. 10 per cent of biotechnology and therefore might represent annual expenditures of around £100 million.

Glossary and Abbreviations

Acidophiles – micro-organisms that show a preference for growth at low pH, *e.g.*, bacteria that grow only at very low pH values, ca. 2.0.

Actinomycetes – members of an order of bacteria in which species are characterised by the formation of branching and/or true filaments.

Activated carbon – a form of carbon prepared to have a very large suface area and suitable for adsorption of organic materials and the support biofilms.

Activated sludge process – an aerobic secondary sewage treatment process using sewage sludge containing active complex populations of aerobic micro-organisms to break down organic matter in sewage.

Adaptation – a structure of behaviour that enhances survival and reproductive potential of an organism in an environment.

Adsorption – a surface phenomenon involving the retention of solid, liquid, or gaseous molecules at an interface.

Aerated pile method – method of composting for the decomposition of organic waste material where the wastes are heaped in separate piles and forced aeration provides oxygen.

Aerobic – having molecular oxygen present; growing in the presence of air.

Algae – a heterogeneous group of eukaryotic, photosynthetic, unicellular and multicellular organisms lacking true tissue differentiation. Includes seaweeds.

Alkanes – straight and branched chain saturated hydrocarbons, constituents of oil and natural gas.

Amensalism – an interactive association between two populations that is detrimental to one while not adversely affecting the other.

Amplification – use of the polymerase chain reaction to produce multiple replications of pieces of genetic material (DNA).

Anaerobes – organisms that grow in the absence of air or oxygen; organisms that do not use molecular oxygen in respiration.

Anaerobic – the absence of oxygen; able to live or grow in the absence of free oxygen.

Anaerobic digester – a secondary sewage treatment facility used for the degradation of sludge and solid waste.

Anoxic – characterised by the absence of oxygen.

Antibody – a complex protein (immunoglobulin) produced in response to an antigen that reacts and binds specifically to form an antigen-antibody complex.

Antigen – a large complex molecule (usually a protein or a carbohydrate) of distinctive shape or functional group that, when introduced into the body of a higher animal, can induce the formation of antibodies which react specifically with it.

Aquifer – a geological formation containing water, such as sub-surface water bodies that supply the water for wells and springs; a permeable layer of rock or soil that holds and transmits water.

Autochthonous – micro-organisms and/or substances indigenous to a given ecosystem; the true inhabitants of an ecosystem; referring to the common microbiota of the body or soil micro-organisms that tend to remain constant despite fluctuations in the quantity of fermentable organic matter.

Autotroph – an organism that obtains its energy from a source other than the oxidation of organic compounds (*i.e.* from inorganic compounds or light) and its carbon from carbon dioxide.

Bacteria – members of a group of diverse and ubiquitous prokaryotic, single-celled organisms; organisms with prokaryotic cells, *i.e.*, cells lacking a nucleus.

Bioaccumulation – accumulation of pollutant residues in the environment.

Bioavailability – the degree of availability of pollutants in contaminated soil or land to biodegradation.

Biocatalyst – an enzyme, used to catalyse a chemical reaction.

Biodegradable – a substance that can be broken down into smaller molecules by the action of biological systems.

Biodegradation – the microbially mediated process of chemical breakdown of a substance to smaller products caused by micro-organisms or their enzymes.

Biodiversity – the variety of different types or species of organisms occuring together in a biological community.

Biofilm – a microbial community occurring on a surface as a microlayer.

Biofilter – a device used for the bioremediation of polluted air consisting of an immobilised microbial community as a biofilm through which the air is passed to detoxify contaminants.

Biogas – gas produced by anaerobic micro-organisms, primarily methane (60-70% plus 30-40% CO_2).

Bioleaching – the use of micro-organisms to transform compounds so that their elements can be extracted from a material when water is filtered through it.

Biological methane potential – a measure of the potential for anaerobic production of methane.

Biological oxygen demand (BOD) – the amount of dissolved oxygen required by aerobic and facultative micro-organisms to stabilise organic matter in sewage or water; also known as biochemical oxygen demand. It is traditionally obtained by measuring the amount of oxygen consumed over 5 days at 20 °C.

Bioluminescence – see Chemiluminescence.

Biomass – all organic matter that derives from the photosynthetic conversion of solar energy; the total mass of living organisms in an ecosystem.

Biopolymer – naturally occurring macromolecules including proteins, polysaccharides and nucleic acids.

Bioreactor – a contained vessel or other structure in which chemical reactions are carried out (usually on an industrial scale), mediated by a biological system, enzymes or cells.

Bioreclamation – use of biological systems to reclaim valuable products from waste streams.

Bioremediation – the use of biological agents to reclaim soils and waters polluted by substances hazardous to human health and/or the environment; it is an extension of biological treatment processes that have traditionally been used to treat wastes in which micro-organisms typically are used to biodegrade environmental pollutants.

Biosensor – an immunological or genetic technique for detecting chemicals or microbial activity, based on the generation of light and/or an electrical signal.

Bioscrubber – a device in which air moves through a fine spray or a microbial suspension in order to remove pollutants from the air.

Biosurfactant – a surface active agent produced by micro-organisms.

Biotower – a device similar to a trickling filter that employs a biofilm through which air is passed to remove pollutants.

Biotrickling filter – a device in which air or off-gases pass counter-current to a liquid flowing over a solid medium supporting a biofilm.

BTEX – the group of chemicals: benzene, toluene, ethylbenzene and xylene. Usually taken together when discussing the bioremediation of contaminated land.

C3 plants – plants having a specific pathway of photosynthetic carbon incorporation which makes them somewhat more efficient at higher temperatures and carbon dioxide concentrations.

Carbonaceous – containing carbon.

Carbon budget – the natural inflows and outflows of carbonaceous materials in a soil.

Cell fusion – a technique of fusing two cells from different species to create one hybrid cell for the purpose of combining some of the genetic characteristics of each original.

Chelating agent – a substance that binds metal ions and also organic molecules.

Chemical oxygen demand (COD) – the amount of oxygen required to oxidise completely the organic matter in a water sample.

Chemiluminescence, bioluminescence – the generation of light by some chemical and biological reactions; particularly by certain bacteria which oxidise proteins called luciferins by an enzyme, luciferase, in the presence of oxygen.

Chemoautotrophs, chemolithotrophs – micro-organisms that obtain their energy from the oxidation of inorganic compounds and their carbon from carbon dioxide.

Chemostat – a bioreactor with identical flows in and out of the reactor to maintain a constant liquid volume, and with generally steady-state conditions with respect to biomass and nutrient concentrations.

Chromogenic – a reaction (or an enzyme whose reaction product) which produces a coloured product.

Chromosome – chemical package of hereditary information (genes), made up of long coiled chains or circles of DNA. Found in the nucleus of cells usually with associated proteins.

Clone – cell or organism identical to an ancestor with respect to genotype and phenotype.

Colloidal – property of a stable mixture of very small particles of one material (solid) in another (liquid). Their properties are usually intermediate between those of a solution and of a suspension.

Co-metabolism – the gratuitous metabolic transformation of a substance by a micro-organism growing on another substrate; the co-metabolised substance is not incorporated into an organism's biomass, and the organism does not derive energy from the transformation of that substance.

Commensalism – an interactive association between two populations of different species living together in which one population benefits from the association, while the other is not affected.

Community – highest biological unit in an ecological hierarchy composed of interacting populations.

Conjugation – the one-way transfer of DNA between bacteria in contact.

Consortium – an interactive association between micro-organisms that generally results in combined metabolic activities.

Culture – the growth (or to grow) cells (*e.g.* micro-organisms, animal or plant cells, etc.) under controlled conditions, for example on a culture plate or flask or in a bioreactor; a single strain of micro-organism.

Dehalogenation (dechlorination, etc.) – the removal of halogen atoms (chlorine, fluorine, etc.) by chemical or biological means.

Denitrification – the formation of gaseous nitrogen or gaseous nitrogen oxides from nitrate or nitrite by micro-organisms.

Deoxyribonucleic acid (DNA) – polymer composed of deoxyribonucleotide units; genetic material of all organisms except RNA viruses.

Desulphurisation – the removal of inorganic or organic sulphur-containing compounds, usually from fossil fuels.

Diastatic enzymes – a group of enzymes present in germinated barley which hydrolyse starch to maltose.

Dioxins – a group of polychlorinated organic compounds often produced by the incomplete combustion of other chlorine-containing materials, noted for their toxicity and extreme recalcitrance to biological degradation.

Domestic sewage – household liquid wastes.

Ecosystem – a functional self-supporting system that includes the organisms in a natural community and their environment.

Effluent – the liquid or gaseous discharge from a process, *e.g.* sewage treatment or industrial production.

Electron acceptor – an agent capable of accepting electrons from (oxidising) a suitable electron donor (oxidisable) substrate.

Enrichment culture – any form of culture in a liquid medium that results in an increase in a given type of organism while minimising the growth of any other organism present.

Enzyme – a protein which catalyses the conversion of a substrate to a product.

Enzyme-linked immunosorbent assay (ELISA) – a technique used for detecting and quantifying specific serum antibodies based upon tagging the antigen-antibody complex with a substrate that can be enzymatically converted to a readily quantifiable product by a specific enzyme.

Eukaryote – an organism (animals, plants, protozoa, fungi and algae but *not* bacteria which are prokaryotes) based on a highly differentiated cell containing a membrane-bound nucleus.

Eutrophication – the enrichment of natural waters with inorganic materials, especially nitrogen and phosphorus compounds, that support the excessive growth of photosynthetic organisms. This may lead to oxygen depletion and the death of other, oxygen-requiring organisms such as fish.

Ex situ – off site, usually used in soil remediation to indicate treatment in which soil is removed to another location for treatment.

Extracellular (exo-) enzyme – an enzyme released by a cell into the medium in which it grows.

Extreme environments – environments characterised by extremes in growth conditions, including temperature, salinity, pH, and water availability, among others.

Extremophile – organisms whose optimum growth is under extreme conditions of temperature, etc.

Fermentation – an anaerobic process where the carbon source is also the electron acceptor (q.v.). Fermentation is used in various industrial processes for the manufacture of products such as alcohols, acids, etc.

Fatty acid – one of a group of aliphatic carboxylic acids.

Floc – a mass of micro-organisms cemented together in a slime produced by certain bacteria, usually found in waste treatment plants.

Flocculant – an agent that causes small particles to aggregate (flocculate).

Fungi – a group of diverse, unicellular and multicellular eukaryotic organisms, lacking chlorophyll, often filamentous and spore-producing.

Gene – the basic unit of heredity, an ordered sequence of nucleotide bases comprising a segment of DNA. A gene contains the sequence of DNA which encodes one polypeptide chain.

Gene expression – the production by a cell of the polypeptide encoded by that gene.

Genetic construct – a segment of DNA, usually a gene or sequence of genes, produced by genetic engineering techniques.

Genetic engineering – the deliberate modification of the genetic properties of an organism by the application of recombinant DNA (q.v.) technology.

Genetic marker – a specific segment of DNA introduced into an organism to allow it to be traced.

Genetic trait – a genetically determined characteristic of an organism.

Genome – the genetic endowment of an organism. When expressed, this will result in the observable characteristics or phenotype.

Groundwater – subsurface water in a terrestrial environment.

Heterotrophs – organisms requiring organic compounds for growth and reproduction, the organic compounds serve as both sources of carbon and energy.

Humus – the organic portion of the soil or organic waste remaining after microbial decomposition.

Hybridisation – a procedure in which single strands of DNA and/or RNA are mixed and subsequently bind to one another. The degree of binding is a measure of the relatedness of the strands. The procedure is used to detect RNA or DNA using suitable probes (q.v.).

Hyperaccumulator plants – species of plants noted for their ability to concentrate relatively high levels of, for example, metals, in their tissues.

Immobilisation – the binding of organisms, cells or enzymes to a substrate such as activated carbon in order to permit the easier separation of reaction products.

Immunoassay – an analytical method that makes use of an antibody which interacts specifically with an antigen (analyte), allowing the quantitation of the target analyte.

Incubation – the maintenance of a culture under regulated conditions, usually of temperature.

Indicator organism – an organism used to identify a particular condition, such as *Escherichia coli* as an indicator of faecal contamination.

Induction – the inducing of a gene to function by a metabolite or external agent (see also repression).

In situ – on site, usually used in soil remediation to mean treatment without moving (digging out) the soil.

Ion exchange – materials, often resins, which permit the replacement of an ion in an incoming stream with one previously bound to the resin. Used as a means of softening water or removing toxic metals and other ionic materials from waste streams.

Lac gene – see Reporter gene.

Landfarming (landtreatment) – the application of toxic organic wastes to soils for the purpose of biodegradation.

Landfill – a site where solid waste is dumped and allowed to decompose; a process in which solid waste containing both inorganic and organic material is deposited and covered with soil.

Leachate – the liquid product of leaching.

194

Leaching – the removal of a soluble compound such as an ore, but also soluble organic compounds, from a solid mixture by washing or percolating.

Lipase – an enzyme which catalyses the breakdown of lipids such as fats.

Lux gene – see Reporter gene.

Matrix – *i)* a rectangular array of elements set out in row and columns to illustrate possible interactions; *ii)* a material in which other discrete materials may be embedded or onto which they may be adsorbed.

Mesophiles – organisms whose optimum growth is in the temperature range of 20 °C-45 °C.

Metabolic potential – resource of degradative capacity in the natural environment.

Methanogens – methane-producing prokaryotes; a group of archaebacteria capable of reducing carbon dioxide or low-molecular-weight fatty acids to produce methane.

Methanophile – an organism requiring methane for reaction or survival.

Methanotrophs – anaerobic micro-organisms utilising methane as sole source of carbon.

Microaerophilic, microaerobic – micro-organisms which will grow in habitats (or the habitats themselves) where oxygen is not strictly absent but is at very low concentrations.

Microbe – synonymous with micro-organism.

Microbial ecology – the field of study that examines the interactions of micro-organisms with their biotic and abiotic surroundings.

Microcosm – a community, often artificial, which is representative of a larger system.

Micro-organisms – microscopic organisms, including algae, bacteria, yeasts, fungi, protozoa, and viruses.

Molecular damage – for example, breakage of nucleic acid chains in a living organism.

Monoclonal antibody – a homogeneous antibody population derived from one specific B-lymphocyte or hybrid cell (see cell fusion).

Morphology – the structure and form of an organism.

Monoculture, pure culture – commercial cultivation of a single variety of plant, or a culture that contains cells of one kind, the progeny of a single cell.

Moulds – any of a group of saprophytic (q.v.) fungi that develop in damp atmospheres on the surface of food, fabrics, etc.

Mutagen – an agent capable of inducing a mutation (a change that alters the sequence or chemistry of bases in the DNA molecule) in the genetic material of an organism.

Mutualism – a stable condition in which two organisms of different species live in close physical association, each organism deriving some benefit from the association; symbiosis.

Nitrification – the process in which ammonia is oxidised to nitrite and nitrite to nitrate; a process primarily carried out by the strictly aerobic, chemolithotrophic bacteria of the family *Nitrobacteraceae*.

Obligate anaerobes – organisms that cannot use molecular oxygen and for which, oxygen can be highly toxic; organisms that grow only under anaerobic conditions, 4, in the absence of air or oxygen; organisms that cannot carry out respiratory metabolism.

Obligate thermophiles – organisms restricted to growth at high temperatures.

Oxygenase – an enzyme involved in the sequential transfer of electrons from a suitable (oxidisable) substrate to oxygen.

Oxygen electrode – a detector designed to measure the concentration of dissolved oxygen in a liquid.

Ozonolysis – the breakdown of organic molecules by ozone usually in conjunction with UV light.

Pasteurisation – the process of heating beverages or foods (usually) to destroy undesirable micro-organisms; limiting the rate of fermentation by the controlled application of heat.

Pathogen – a disease-producing agent, usually restricted to a living agent such as a bacterium or a virus.

Pesticide – chemical product used to destroy pests (*e.g.* insecticide, herbicide, etc.).

pH – the symbol used to express the hydrogen ion concentration (acidity), signifying the logarithm to the base 10 of the reciprocal of the hydrogen ion concentration.

Phage – a category of virus which specifically attacks micro-organisms.

Phenotype – the characteristics of an organism that result from the interaction of its genetic constitution with the environment.

Photosynthesis – the synthesis by plants of organic compounds from carbon dioxide and water using light energy absorbed by chlorophyll.

Phytoplankton – passively floating or weakly motile photosynthetic aquatic organisms, primarily cyanobacteria and algae.

Phytotoxic – substance toxic to plants.

Plasmid – an extrachromosomal, self-replicating, circular segment of DNA; plasmids and some viruses are used as vectors (*q.v.*) for cloning DNA in bacterial host cells.

Polychlorinated biphenyls (PCBs) – a group of compounds once of industrial importance but no longer used because of their toxic qualities and extreme recalcitrance to biodegradation.

Polyclonal antibodies – a heterogeneous population of different antibodies varying in both antigenic specificity and affinity.

Polycyclic aromatic hydrocarbons – a group of carcinogenic and generally recalcitrant compounds often found as pollutants in soils and dredging wastes.

Polymerase chain reaction (PCR) – a technique using the enzyme polymerase to produce many copies of a nucleotide sequence.

Polyphosphate – compound containing a chain of phosphate groups used as a chelating agent (*q.v.*) in detergents.

Primary sewage treatment – removal of gross solids.

Probe – a device inserted into a reactor to measure levels of chosen parameters. See oxygen electrode, transducer.

Probe (DNA or RNA) – a specific DNA or RNA sequence used to detect complementary sequences among nucleic acid molecules. Can be used as very sensitive biological detector.

Prokaryote – an organism (*e.g.* bacterium) having a cell lacking a nucleus.

Promoter – a DNA sequence whose function is to direct the expression of a gene under its control.

Protein engineering – the generation of proteins (enzymes) with subtly modified structures thus conferring new properties such as changed catalytic specificity or thermal stability.

Protozoa – diverse eukaryotic, typically unicellular, non-photosynthetic micro-organisms generally lacking a rigid cell wall; often ''grazers'' – predators that consume bacteria indiscriminately.

Radioimmunoassay (RIA) – an immunoassay method that involves competition between radioactively labelled and unlabelled antigen for binding sites on antibody molecules.

Recalcitrant – a term applied to pollutants which are not biodegradable or only biodegradable with difficulty.

Recombinant DNA (r-DNA) – a DNA molecule formed by joining DNA segments from two or more sources.

Replication – the copying of a segment of DNA or RNA.

Reporter gene – a gene whose expression is easily detected and that can be used to track the transcription of other genes, *e.g.* the *lux* gene, which codes for light production, and the lac gene which encodes a specific enzyme, and can be used to detect the expression of other genes.

Repression – the prevention of the expression of a gene (*q.v.*) by a metabolite or external agent.

Respiration – aerobic metabolic process utilising oxygen as the ultimate electron acceptor.

Retention time – period during which micro-organisms or other materials are retained in a bioreactor.

Rhizosphere – an ecological niche that comprises the surfaces of plant roots and the region of the surrounding soil in which the microbial populations are affected by the presence of the roots.

Saprophytic – living and growing on dead organic matter.

Saturation kinetics – kinetics of metabolic reactions predicated on saturated concentrations of substrates.

Secondary sewage treatment – the treatment of the liquid portion of sewage containing dissolved organic matter, using micro-organisms to degrade the organic matter that is mineralised or converted to removable solids.

Sequestering – to remove or separate, usually by the use of a chemical agent such as a chelating or flocculating agent, but equally by biomass or living organisms.

Sewage treatment – the treatment of sewage to reduce its biological oxygen demand and to inactivate the pathogenic micro-organisms present.

Sour gas – hydrogen sulphide, usually as a contaminant of natural gas.

Strain – a group of organisms within a species or variety distinguished by one or more minor characteristics; a variety of bacterium or fungus used for culturing.

Substrate – *i)* the chemical substance acted upon by an enzyme; *ii)* a base support on which other material is deposited, adsorbed or immobilised.

Suicide gene – a gene whose expression causes the destruction of the organism itself. Suicide genes may be incorporated in GMOs which may be released into the environment.

Symbiosis – a partnership between two organisms for their mutual benefit.

Taxon – a group of organisms based on similarities of structure or origin.

Tertiary sewage treatment – removal of suspended solids following secondary sewage treatment (*q.v.*).

Thermophiles – organisms having an optimum growth temperature above 45 °C.

Toxic shock – temporary or permanent collapse of a digestion process caused by a sudden excess of pollutant.

Transducer – a device which converts an action (a movement or change in concentration, for example) into a (usually) electrical signal.

Transduction – transfer of DNA from a donor to a recipient strain utilising a virus as a vector.

Transformation – uptake of naked DNA by a competent recipient strain.

Transgenic – organisms into which DNA from another species are introduced by, for example, microinjection or retroviral infection.

Transposon – a fragment of DNA which can occur in a number of different chromosomal or extrachromosomal positions.

Trickling filter system – a simple, film-flow aerobic sewage treatment system; the sewage is distributed over a porous bed coated with bacterial growth that mineralises the dissolved organic nutrients.

UASB – upflow anaerobic sludge blanket. Abbreviation stands for one type of anaerobic waste water treatment installation producing biogas.

Ultraviolet (UV) – short wavelength electromagnetic radiation in the range 100-400 nm. Often subdivided into UV-A and UV-B (280-320 nm).

Vector – an agent of transmission; for example, a DNA vector is a self-replicating segment of DNA that transmits genetic information from one cell or organism to another.

Virus – a noncellular entity that consists minimally of protein and nucleic acid and that can replicate only after entry into specific types of living cells, and then only by making use of the cell's own systems.

Voidage – the air space within a soil or within the solid bed of a biofilter.

Xenobiotic – a synthetic product not formed by natural biosynthetic processes; a foreign substance or poison.

Xerophyte – a plant adapted to living in a dry habitat (or the habitat itself).

Yeast – a category of fungi (*q.v.*) defined in terms of morphological and physiological criteria; typically unicellular, saprophytic (*q.v.*) organisms that characteristically ferment a range of carbohydrates; commercially used for brewing, wine making and breadmaking.

Abbreviations

AGB	above-ground bioreactor
BMP	biological methane potential
BOD	biological (or biochemical) oxygen demand
BTEX	benzene, toluene, ethylbenzene and xylene
COD	chemical oxygen demand
DNA	deoxyribonucleic acid
ELISA	enzyme-linked immunosorbent assay
GEM	genetically engineered micro-organism
GMO	genetically modified organism
PAH	polycyclic-aromatic hydrocarbon
PCB	polychlorinated biphenyl
PCP	pentachlorophenol

PCR	polymerase chain reaction
ppb	parts per billion
ppm	parts per million
r-DNA	recombinant DNA
RIA	radioimmunoassay
RNA	ribonucleic acid
TCE	trichloroethylene
UASB	upflow anaerobic sludge blanket
UN	United Nations
US EPA	United States Environmental Protection Agency
VOC	volatile organic compound
WHO	World Health Organisation

Ad Hoc Group of Government Experts on Biotechnology for a Clean Environment: Participants

The following national experts attended one or more of the four meetings of the Ad Hoc Group on Biotechnology for a Clean Environment (11-12 July 1991, 27-28 April 1992, 11-12 January 1993, 14-15 October 1993).

Chairman

Dr. Mike GRIFFITHS
Mike Griffiths Associates
Woking, Surrey
United Kingdom

Austria

Prof. R. BRAUN
Assistant Professor
Department of Applied Microbiology
University of Agriculture & Forestry
Vienna

Dr. Manfred SCHNEIDER
Federal Environment Agency
Vienna

Belgium

Mr. Jan de BRABANDERE
Programme Manager
Services de Programmation de la Politique
   Scientifique
Brussels

Prof. Willy VERSTRAETE
Laboratory of Microbial Ecology
Ghent

Denmark

Prof. Ebba LUND
Royal Veterinary & Agriculture University
Frederiksberg

Dr. V. RASMUSSEN
Royal Veterinary & Agriculture
University of Copenhagen
Department of Veterinary, Virology & Immunology
Copenhagen

Germany

Dr. B. BOSSOW-BERKE
Federal Ministry for Research and Technology
   (BMFT)
Bonn

Dr. A. GOERDELER
Federal Ministry for Research and
   Technology (BMFT)
Bonn

Dr. Barbel HUSING
Fraunhofer Institute for Systems &
   Innovation Research
Karlsruhe

Dr. MÜLLER-KUHRT
Berlin

Dr. Thomas REISS
Fraunhofer Institute for Systems &
   Innovation Research
Karlsruhe

Hungary

Mr. A. ZSIGMOND
National Office for Technological Development
Department of International Relations
Budapest

Italy

Dr. Gabriella LEVI
ENEA – Area Energia & Ambiente
Rome

Prof. Andrea ROBERTIELLO
Eniricerche
Environmental Protection Department
Rome

Japan

Mr. Osamu CHISAKI
Biochemical Industry Division
Ministry of International Trade & Industry
Tokyo

Mr. Masahiro HASHIMOTO
Biochemical-Industry Division
Basic Industries Bureau,
Ministry of International Trade & Industry
Tokyo

Prof. Toyohiko HAYAKAWA
Chiba Institute of Technology
Chiba-ken

Mr. Ken-ichi HAYASHI
Society for Techno-Innovation of
Agriculture, Forestry & Fisheries
Tokyo

Mr. Akio IKEMORI
Biochemical – Industry Division
Basic Industries Bureau
Ministry of International Trade & Industry
Tokyo

Mr. Ryoichi KIKUCHI
Environmental Health Science Laboratory
Sumitomo Chemical Co. Ltd
Osaka

Dr. R. KURANE
Fermentation Research Institute
Agency of Industrial Science & Technology
Ministry of International Trade & Industry
Ibaragi

Mr. M. MASUDA
Bio-chemical Industries Division
Ministry of International Trade & Industry
Tokyo

Mr. Eiichi TAKESHITA
Mitsubishi Petrochemical Co. Ltd
Tokyo

Mr. Satoshi TANIKAWA
International Trade & Industry
Inspection Institute,
Ministry of International Trade & Industry
Tokyo

Mr. Jihei YODA
JBA
Tokyo

The Netherlands

Dr. H. DODDEMA
TNO – Institute Environmental Sciences

Dept. of Environmental Biotechnology
Delft

Dr. Paul HESSELINK
TNO – Institute Environmental Sciences
Dept. Environmental Biotechnology
Delft

Mr. Joop MAASAKKERS
Ministry of Economic Affairs
Biotechnology Policy Unit
The Hague

Spain

Prof. Victor de LORENZO
Centro de Investigacione Biologicas
Madrid

Dr. Eladio MONTOYA
Secretaria General del Plan Nacional de I&D
Comision Interministerial de Ciencia
   y Tecnologia
Madrid

Sweden

Dr. Gustaf BRUNIUS
The Swedish Recombinant DNA Advisory
Committee
Stockholm

Switzerland

Prof. A. FIECHTER
ETH Zurich
Institute of Biotechnology
Zurich

United Kingdom

Dr. Julie RANKIN
Biotech Unit RM B357
Department of the Environment
London

Dr. P. VAUGHAN
Biotechnology Unit
Department of Trade & Industry
Teddington, Middlesex

United States

Prof. Ronald ATLAS
Department of Biology
University of Louisville
Louisville

Mr. Tom BAUGH
Environmental Protection Agency
Office of Environmental Engineering
   & Technology Demonstration
Washington DC

Mr. George DAVATELIS
Office of Ecology, Health & Conservation
Bureau of Oceans & International
Environmental & Scientific Affairs

Department of State
Washington DC

Dr. A. LINDSEY
Office of Environmental Engineering &
    Technological Demonstration
Environmental Protection Agency
Washington DC

Dr. Robert MENZER
US Environmental Protection Agency
Environmental Research Laboratory
Gulf Breeze, Florida

Mr. Michael SCHECHTMAN
Animal and Plant Health Inspection
Service, BEEP
Department of Agriculture
Hyattsville, MD

Commission of the European Communities

Mr. Ioannis ECONOMIDIS
Commission of the European Communities
Directorate General for Science, Research
    & Development

Brussels
Belgium

Mr. Holger PEDERSEN
Commission of the European Communities
Brussels
Belgium

OECD Secretariat

Directorate for Science, Technology & Industry
    Mr. Yoshitaka ANDO
    Mr. Mark CANTLEY
    Dr. Seizo SUMIDA
    Dr. Salomon WALD

Environment Directorate
    Mr. Peter KEARNS

Development Centre
    Ms C. BRENNER

MAIN SALES OUTLETS OF OECD PUBLICATIONS
PRINCIPAUX POINTS DE VENTE DES PUBLICATIONS DE L'OCDE

ARGENTINA – ARGENTINE
Carlos Hirsch S.R.L.
Galería Güemes, Florida 165, 4° Piso
1333 Buenos Aires   Tel. (1) 331.1787 y 331.2391
Telefax: (1) 331.1787

AUSTRALIA – AUSTRALIE
D.A. Information Services
648 Whitehorse Road, P.O.B 163
Mitcham, Victoria 3132   Tel. (03) 873.4411
Telefax: (03) 873.5679

AUSTRIA – AUTRICHE
Gerold & Co.
Graben 31
Wien I   Tel. (0222) 533.50.14

BELGIUM – BELGIQUE
Jean De Lannoy
Avenue du Roi 202
B-1060 Bruxelles   Tel. (02) 538.51.69/538.08.41
Telefax: (02) 538.08.41

CANADA
Renouf Publishing Company Ltd.
1294 Algoma Road
Ottawa, ON K1B 3W8   Tel. (613) 741.4333
Telefax: (613) 741.5439
Stores:
61 Sparks Street
Ottawa, ON K1P 5R1   Tel. (613) 238.8985
211 Yonge Street
Toronto, ON M5B 1M4   Tel. (416) 363.3171
Telefax: (416)363.59.63
Les Éditions La Liberté Inc.
3020 Chemin Sainte-Foy
Sainte-Foy, PQ G1X 3V6   Tel. (418) 658.3763
Telefax: (418) 658.3763

Federal Publications Inc.
165 University Avenue, Suite 701
Toronto, ON M5H 3B8   Tel. (416) 860.1611
Telefax: (416) 860.1608
Les Publications Fédérales
1185 Université
Montréal, QC H3B 3A7   Tel. (514) 954.1633
Telefax : (514) 954.1635

CHINA – CHINE
China National Publications Import
Export Corporation (CNPIEC)
16 Gongti E. Road, Chaoyang District
P.O. Box 88 or 50
Beijing 100704 PR   Tel. (01) 506.6688
Telefax: (01) 506.3101

DENMARK – DANEMARK
Munksgaard Book and Subscription Service
35, Nørre Søgade, P.O. Box 2148
DK-1016 København K   Tel. (33) 12.85.70
Telefax: (33) 12.93.87

FINLAND – FINLANDE
Akateeminen Kirjakauppa
Keskuskatu 1, P.O. Box 128
00100 Helsinki
Subscription Services/Agence d'abonnements :
P.O. Box 23
00371 Helsinki   Tel. (358 0) 12141
Telefax: (358 0) 121.4450

FRANCE
OECD/OCDE
Mail Orders/Commandes par correspondance:
2, rue André-Pascal
75775 Paris Cedex 16   Tel. (33-1) 45.24.82.00
Telefax: (33-1) 49.10.42.76
Telex: 640048 OCDE

OECD Bookshop/Librairie de l'OCDE :
33, rue Octave-Feuillet
75016 Paris   Tel. (33-1) 45.24.81.67
(33-1) 45.24.81.81
Documentation Française
29, quai Voltaire
75007 Paris   Tel. 40.15.70.00
Gibert Jeune (Droit-Économie)
6, place Saint-Michel
75006 Paris   Tel. 43.25.91.19
Librairie du Commerce International
10, avenue d'Iéna
75016 Paris   Tel. 40.73.34.60
Librairie Dunod
Université Paris-Dauphine
Place du Maréchal de Lattre de Tassigny
75016 Paris   Tel. (1) 44.05.40.13
Librairie Lavoisier
11, rue Lavoisier
75008 Paris   Tel. 42.65.39.95
Librairie L.G.D.J. - Montchrestien
20, rue Soufflot
75005 Paris   Tel. 46.33.89.85
Librairie des Sciences Politiques
30, rue Saint-Guillaume
75007 Paris   Tel. 45.48.36.02
P.U.F.
49, boulevard Saint-Michel
75005 Paris   Tel. 43.25.83.40
Librairie de l'Université
12a, rue Nazareth
13100 Aix-en-Provence   Tel. (16) 42.26.18.08
Documentation Française
165, rue Garibaldi
69003 Lyon   Tel. (16) 78.63.32.23
Librairie Decitre
29, place Bellecour
69002 Lyon   Tel. (16) 72.40.54.54

GERMANY – ALLEMAGNE
OECD Publications and Information Centre
August-Bebel-Allee 6
D-53175 Bonn   Tel. (0228) 959.120
Telefax: (0228) 959.12.17

GREECE – GRÈCE
Librairie Kauffmann
Mavrokordatou 9
106 78 Athens   Tel. (01) 32.55.321
Telefax: (01) 36.33.967

HONG-KONG
Swindon Book Co. Ltd.
13–15 Lock Road
Kowloon, Hong Kong   Tel. 366.80.31
Telefax: 739.49.75

HUNGARY – HONGRIE
Euro Info Service
Margitsziget, Európa Ház
1138 Budapest   Tel. (1) 111.62.16
Telefax : (1) 111.60.61

ICELAND – ISLANDE
Mál Mog Menning
Laugavegi 18, Pósthólf 392
121 Reykjavik   Tel. 162.35.23

INDIA – INDE
Oxford Book and Stationery Co.
Scindia House
New Delhi 110001   Tel.(11) 331.5896/5308
Telefax: (11) 332.5993
17 Park Street
Calcutta 700016   Tel. 240832

INDONESIA – INDONÉSIE
Pdii-Lipi
P.O. Box 269/JKSMG/88
Jakarta 12790   Tel. 583467
Telex: 62 875

ISRAEL
Praedicta
5 Shatner Street
P.O. Box 34030
Jerusalem 91430   Tel. (2) 52.84.90/1/2
Telefax: (2) 52.84.93
R.O.Y.
P.O. Box 13056
Tel Aviv 61130   Tél. (3) 49.61.08
Telefax (3) 544.60.39

ITALY – ITALIE
Libreria Commissionaria Sansoni
Via Duca di Calabria 1/1
50125 Firenze   Tel. (055) 64.54.15
Telefax: (055) 64.12.57
Via Bartolini 29
20155 Milano   Tel. (02) 36.50.83
Editrice e Libreria Herder
Piazza Montecitorio 120
00186 Roma   Tel. 679.46.28
Telefax: 678.47.51
Libreria Hoepli
Via Hoepli 5
20121 Milano   Tel. (02) 86.54.46
Telefax: (02) 805.28.86
Libreria Scientifica
Dott. Lucio de Biasio 'Aeiou'
Via Coronelli, 6
20146 Milano   Tel. (02) 48.95.45.52
Telefax: (02) 48.95.45.48

JAPAN – JAPON
OECD Publications and Information Centre
Landic Akasaka Building
2-3-4 Akasaka, Minato-ku
Tokyo 107   Tel. (81.3) 3586.2016
Telefax: (81.3) 3584.7929

KOREA – CORÉE
Kyobo Book Centre Co. Ltd.
P.O. Box 1658, Kwang Hwa Moon
Seoul   Tel. 730.78.91
Telefax: 735.00.30

MALAYSIA – MALAISIE
Co-operative Bookshop Ltd.
University of Malaya
P.O. Box 1127, Jalan Pantai Baru
59700 Kuala Lumpur
Malaysia   Tel. 756.5000/756.5425
Telefax: 757.3661

MEXICO – MEXIQUE
Revistas y Periodicos Internacionales S.A. de C.V.
Florencia 57 - 1004
Mexico, D.F. 06600   Tel. 207.81.00
Telefax : 208.39.79

NETHERLANDS – PAYS-BAS
SDU Uitgeverij Plantijnstraat
Externe Fondsen
Postbus 20014
2500 EA's-Gravenhage   Tel. (070) 37.89.880
Voor bestellingen: Telefax: (070) 34.75.778

NEW ZEALAND
NOUVELLE-ZÉLANDE
Legislation Services
P.O. Box 12418
Thorndon, Wellington   Tel. (04) 496.5652
Telefax: (04) 496.5698

NORWAY – NORVÈGE
Narvesen Info Center – NIC
Bertrand Narvesens vei 2
P.O. Box 6125 Etterstad
0602 Oslo 6   Tel. (022) 57.33.00
                        Telefax: (022) 68.19.01

PAKISTAN
Mirza Book Agency
65 Shahrah Quaid-E-Azam
Lahore 54000   Tel. (42) 353.601
                        Telefax: (42) 231.730

PHILIPPINE – PHILIPPINES
International Book Center
5th Floor, Filipinas Life Bldg.
Ayala Avenue
Metro Manila   Tel. 81.96.76
                        Telex 23312 RHP PH

PORTUGAL
Livraria Portugal
Rua do Carmo 70-74
Apart. 2681
1200 Lisboa   Tel.: (01) 347.49.82/5
                        Telefax: (01) 347.02.64

SINGAPORE – SINGAPOUR
Gower Asia Pacific Pte Ltd.
Golden Wheel Building
41, Kallang Pudding Road, No. 04-03
Singapore 1334   Tel. 741.5166
                        Telefax: 742.9356

SPAIN – ESPAGNE
Mundi-Prensa Libros S.A.
Castelló 37, Apartado 1223
Madrid 28001   Tel. (91) 431.33.99
                        Telefax: (91) 575.39.98

Libreria Internacional AEDOS
Consejo de Ciento 391
08009 – Barcelona   Tel. (93) 488.30.09
                        Telefax: (93) 487.76.59
Llibreria de la Generalitat
Palau Moja
Rambla dels Estudis, 118
08002 – Barcelona
          (Subscripcions) Tel. (93) 318.80.12
          (Publicacions) Tel. (93) 302.67.23
                        Telefax: (93) 412.18.54

SRI LANKA
Centre for Policy Research
c/o Colombo Agencies Ltd.
No. 300-304, Galle Road
Colombo 3   Tel. (1) 574240, 573551-2
                        Telefax: (1) 575394, 510711

SWEDEN – SUÈDE
Fritzes Information Center
Box 16356
Regeringsgatan 12
106 47 Stockholm   Tel. (08) 690.90.90
                        Telefax: (08) 20.50.21
Subscription Agency/Agence d'abonnements :
Wennergren-Williams Info AB
P.O. Box 1305
171 25 Solna   Tel. (08) 705.97.50
                        Téléfax : (08) 27.00.71

SWITZERLAND – SUISSE
Maditec S.A. (Books and Periodicals - Livres
et périodiques)
Chemin des Palettes 4
Case postale 266
1020 Renens   Tel. (021) 635.08.65
                        Telefax: (021) 635.07.80

Librairie Payot S.A.
4, place Pépinet
CP 3212
1002 Lausanne   Tel. (021) 341.33.48
                        Telefax: (021) 341.33.45

Librairie Unilivres
6, rue de Candolle
1205 Genève   Tel. (022) 320.26.23
                        Telefax: (022) 329.73.18

Subscription Agency/Agence d'abonnements :
Dynapresse Marketing S.A.
38 avenue Vibert
1227 Carouge   Tel.: (022) 308.07.89
                        Telefax : (022) 308.07.99

See also – Voir aussi :
OECD Publications and Information Centre
August-Bebel-Allee 6
D-53175 Bonn (Germany) Tel. (0228) 959.120
                        Telefax: (0228) 959.12.17

TAIWAN – FORMOSE
Good Faith Worldwide Int'l. Co. Ltd.
9th Floor, No. 118, Sec. 2
Chung Hsiao E. Road
Taipei   Tel. (02) 391.7396/391.7397
                        Telefax: (02) 394.9176

THAILAND – THAÏLANDE
Suksit Siam Co. Ltd.
113, 115 Fuang Nakhon Rd.
Opp. Wat Rajbopith
Bangkok 10200   Tel. (662) 225.9531/2
                        Telefax: (662) 222.5188

TURKEY – TURQUIE
Kültür Yayinlari Is-Türk Ltd. Sti.
Atatürk Bulvari No. 191/Kat 13
Kavaklidere/Ankara   Tel. 428.11.40 Ext. 2458
Dolmabahce Cad. No. 29
Besiktas/Istanbul   Tel. 260.71.88
                        Telex: 43482B

UNITED KINGDOM – ROYAUME-UNI
HMSO
Gen. enquiries   Tel. (071) 873 0011
Postal orders only:
P.O. Box 276, London SW8 5DT
Personal Callers HMSO Bookshop
49 High Holborn, London WC1V 6HB
                        Telefax: (071) 873 8200
Branches at: Belfast, Birmingham, Bristol, Edin-
burgh, Manchester

UNITED STATES – ÉTATS-UNIS
OECD Publications and Information Centre
2001 L Street N.W., Suite 700
Washington, D.C. 20036-4910 Tel. (202) 785.6323
                        Telefax: (202) 785.0350

VENEZUELA
Libreria del Este
Avda F. Miranda 52, Aptdo. 60337
Edificio Galipán
Caracas 106   Tel. 951.1705/951.2307/951.1297
                        Telegram: Libreste Caracas

Subscription to OECD periodicals may also be
placed through main subscription agencies.

Les abonnements aux publications périodiques de
l'OCDE peuvent être souscrits auprès des
principales agences d'abonnement.

Orders and inquiries from countries where Distribu-
tors have not yet been appointed should be sent to:
OECD Publications Service, 2 rue André-Pascal,
75775 Paris Cedex 16, France.

Les commandes provenant de pays où l'OCDE n'a
pas encore désigné de distributeur devraient être
adressées à : OCDE, Service des Publications,
2, rue André-Pascal, 75775 Paris Cedex 16, France.

9-1994

OECD PUBLICATIONS, 2 rue André-Pascal, 75775 PARIS CEDEX 16
PRINTED IN FRANCE
(93 94 05 1) ISBN 92-64-14257-6 - No. 47091 1994